THE ART OF WAR IN BUSINESS TECHNOLOGY

SIMPLE TACTICS FOR CONQUERING IT & CYBERSECURITY CHALLENGES ON THE SMALL BUSINESS BATTLEFIELD

A-JAY ORR

www.ajayorr.com

Publishing services provided by **Archangel Ink**

ISBN: 978-1-950043-35-4

CONTENTS

INTRODUCTION

Have you ever had the nagging sense that something isn't quite right with your business, but you can't put your finger on what's wrong? You might be tempted to blow it off or downplay it. "Business is good. Our teams are thriving. We're expanding." But just because things are going in the right direction doesn't mean you should settle. If you're ready to lean into that desire for more, you've come to the right place. I'm here to tell you that you can optimize your business, and I want to give you the tools you need to do it.

My name is A-Jay Orr, and I am the CEO and founder of Simple Plan IT. When I was seventeen, I was taught seven principles: loyalty, duty, respect, selfless service, honor, integrity, and personal courage. If you're familiar with the Army, you probably recognize those principles as our Army values. At seventeen I had no idea that those values would lead me to starting my own business. There was a long stretch after the military when I didn't know where I was going. In the span of twelve months, I lived in three different states and bounced around from job to job, just trying to find my way. I worked fast food, I sold retail, I was a bartender,

I was even a traveling car salesman for a stint. My life is a testament to the truth that the road to success is never a straight path.

There are plenty of twists and turns and ups and downs in this thing called life. But if you keep pushing through the hard times and commit to trying to do the right thing, then eventually you will find your path. At least, that's how it worked out for me. You see, I never had a plan or the desire to start a business. Things like money, power, freedom—all of the things that typically motivate people to start a business—didn't interest me. I was driven by one thing: the challenge.

Fifteen years after my military career, I found myself working for a small technology firm selling technology support services to other businesses. I had never done anything like it before. But what I lacked in skill, I made up for with hard work, refusal to quit, fast learning, and problem-solving. At the time, I didn't realize that taking that job was the first step to living out my calling. I worked with companies of all sizes and in multiple industries. Every client was different and presented their own unique set of issues and circumstances. I soon found out that, in the world of technology, there were countless solutions to every problem. Once I had that revelation, it became a game to me. I wanted to discover all of the options and to study them to see which one made the most sense. It was like playing chess: there were countless moves I *could* make, but only one best move. Finding the best move became my obsession. I was determined to find not only the best solution to the problem the client knew they had, but also the best solution to the problems they had not even considered.

Anchored by my Army values, my obsession led me to the path of entrepreneurship. As I was out in the market cultivating opportunities, I occasionally came across an opportunity where the services that I represented were not the best solution. Rather than selling the prospect on an inferior solution, I decided to point them toward the better option. I even provided them with a contact at the company that could best service their needs. All I asked in return was that they share my name with anyone they met who was having technology issues.

One day I received a call from one of the prospects that I had pointed in another direction. He said, "A-Jay, I appreciate and trust your recommendation, but I met with your guy and didn't understand a damn thing he was saying. Since you understand what we're trying to accomplish, do you mind sitting in on our next meeting to make sure we're getting what we need?" Knowing it wouldn't take long, I agreed to be there, and I made sure that everyone was on the same page. I did the exact same thing a few more times and then, one day, I sat in on a meeting for John.

John's situation was no different than the others except that at the end of the meeting, John gave me an envelope with a check inside. I repeatedly told him he didn't owe me anything and that if he wanted to do something for me, he could send referrals my way. But John wouldn't let me leave without the envelope. I remember it like it was yesterday. I walked out of his building that night into what seemed like a monsoon. The wind blew so hard, the rain fell sideways. I ran to my car, but I was soaked nonetheless. I settled into my seat, took the envelope out of my pocket, and opened it to find a $500 check. My mind was completely blown. I had just been given $500 to sit in on a forty-five minute meeting and be a translator. So the next time I was asked to sit in on a meeting, I asked the prospect if they'd be willing to pay for my time. To my surprise, they said sure and asked what my rate was. I sheepishly said $100, to which they happily agreed. Looking back, I didn't even realize I could charge the $500 fee John had paid me. Like I said, I wasn't planning on starting a business.

As time passed, I developed a reputation, and people sought me out to help them determine the best technology solutions for their business. So, I decided to fill that gap and become a resource for them. My initial goal was simple: I would sit beside them during their vendor meetings to ensure the technology solutions they were putting in place aligned with their strategic goals for the business. I became the trusted advisor for all things technology related for these businesses. As their trusted advisor, my duty was to provide them with simple plans to manage their IT and digital risks.

In 2012, Simple Plan IT was born. Today it is still guided by the principles the Army taught me all those years ago. We have been fortunate enough to help businesses of all sizes in seven different countries to stay secure and to strategically use technology to increase the efficiency and revenue of their business.

I wish I could tell you that my path to business success was straight and easy, but it wasn't. Like the dark and rainy or snowy nights I walked home alone after working a twelve-hour bartending shift when I couldn't afford a car, the journey of a business leader can be cold and lonely. But I am here to tell you that it doesn't have to be. Over the years of helping business leaders maximize the use of technology in their organizations, I realized their businesses had begun to change. Because we were able to reduce bottlenecks and increase efficiency, their employees were able to get more done in a day. With better performance came better attitudes and less stress. I have even had business owners tell me they started enjoying their work again, that the business growth had allowed them to experience the lifestyle they had been working so hard to achieve.

What This Book Is ... and Is Not

As much as I would love for you to read this book, I value your time even more. I don't want you to waste your time on this book if it's not the right fit for you.

This is not a get-rich-quick book. If you're looking

for fast, easy strategies that will make you hundreds of thousands of dollars within the next thirty days, this is ***NOT*** the book for you. If you're just starting your business journey and don't have clients or a steady stream of income, this book is ***NOT*** for you.

This book is for the motivated business leader. The one who is constantly looking for ways to increase their efficiency so they can better service their clients, reduce their stress, and increase their revenue. The one who understands that their business is only as strong as their weakest link and who is committed to building them up.

So, if you're still reading, I want to extend a warm welcome to you. Now that we know you're a motivated leader who likes to take action, I have a question for you:

Why do you want to optimize your business?[1]

For me, and many other business leaders I've had the pleasure to work with, the answer was simple. It came down to one word: time. Time is the one resource we all have that we cannot make more of. Once it's gone, it's gone. There's no getting it back. And if you're like us and want to accomplish many things in your life, you realize you don't have time to waste.

1 Go ahead. Write down your answer. You'll probably want to take notes as you read through the book.

What You Can Expect from This Book

What I'm about to share with you are the tactics and strategies I've used to help countless organizations increase their bottom line, optimize their operations, and enhance the stability of their organization.

When I first had the idea to write this book, I wanted it to be straightforward, easy to understand, and simple to implement. The more I thought about my desired outcome, the more I was reminded of the *U.S. Army Tactics Field Manual.* At its core, the manual provides military leaders with a framework that allows them to maximize the use of their resources in order to win battles. In business, much like in a military battle, you are constantly trying to increase your leverage, gain more ground, and protect the position you have gained.

I have one simple goal with this book: to give you the tools you need to maximize the use of technology in your business and increase your revenue so you can live the lifestyle you desire. To accomplish this goal, the book focuses on the following three areas:

Income

If this journey is going to have a successful outcome, you must first start by being honest with yourself. In order to have the business you desire and live the life you dream of (which should absolutely be your goal), you need money. It doesn't matter if you're aiming to build the next Amazon or a nonprofit that will end homelessness, you'll need money to do it. And depending on the size of your dreams,[2] you may need to increase what you are currently generating by as much as ten times.

Understanding how important revenue and cash-flow is to a business, I wanted to focus on that early in the book. In chapters 2–4, I show you the strategies and tactics I use to help business leaders increase their bottom line.

Efficiency

Like the military, your business has a lot of moving parts. You have sales, marketing, new client onboarding, product/service delivery, customer satisfaction, customer retention, and product/service development. You may not realize that each of these business functions is dependent on the others. In order for your business to run optimally, all of those processes need to work together seamlessly. To scale your business (which should also be part of your goal), you need to have systems in place that allow you to easily replicate your desired outcomes. In chapters 5–8, I share with you the simple approach I use to increase the efficiency and stability of businesses I work with. This approach has allowed them to scale their business with ease.

2 BONUS TIP: if your dreams do not intimidate you, then you are not dreaming big enough.

Defenses

The biggest threat to your success and the success of your business is the threat you don't know exists. In the digital information age we live in, we face threats of many shapes and sizes, and they are constantly evolving. Your competitors are always seeking to figure out what you're doing so they can gain an advantage in the market. Not only are they targeting your customers, but they are also targeting your employees. Now you must watch for disgruntled employees that will steal your intellectual property (e.g., internal checklists, frameworks, client lists, etc.). If that weren't bad enough, you also have to worry about the digital and socially engineered attacks that hackers and scammers are using to steal from you.

The good news is every business is facing these challenges. Cyberattacks are constantly evolving, but our approach to cybersecurity remains the same. Sure, having firewalls and antivirus and intrusion-detection software is important. But the fact that cybercrime is a trillion-dollar industry just goes to show it's not enough. So how do you defend yourself against an enemy that is constantly changing? I will show you how in chapters 9–12.

How to Apply These Tools to Your Personal Life

While this book is designed to help you optimize the performance of your business, I want to show you how these same principles can apply to the rest of your life. At the end of each chapter, you'll find an example of how you can apply the lessons covered in that chapter to your personal life. As I've said, this book aims to support you to live the life that you desire. That could mean spending more time with your family, traveling more, buying expensive toys, or simply taking time off to be one with nature. Regardless of what your dream life looks like, I want to show you how these simple principles can help you achieve success in all aspects of your life.

A Note about Guiding Principles

As you know by now, the Army had a profound impact on me. I attribute my success in life, and in business, to the values they taught me at a young age. Therefore, you are going to see a lot of military references in this book.

While some of the things I share may be foreign to you, that's okay. Trust me when I say you're not alone. Countless others started exactly where you are now and have found the success and business stability that you desire. If you follow the blueprint I lay out in this book, you will have all the tactics and strategies you need to optimize your business and, quite possibly, optimize your life.

Thank you for joining me on this journey! Without any further ado … let's get started.

PERSONAL LIFE APPLICATION

When I first started my career in project management, I always started with the end in mind. I would get really clear on the details that the client wanted in the final solution. Once I had a clear understanding of their desired outcome, creating the path to get there was easy.

One day it hit me that I should apply that same strategy to my personal life. This subtle shift allowed me more enjoyment in life. Life is hard; it's challenging; it can wear you down. Especially if you are not seeing results from the work you put in. In my case, I was having business success, but I lacked joy in my personal life. It wasn't until I got clear on what I wanted in my personal life that things began to change for me.

So for our very first *Personal Life Application* lesson in this book, I want you to write down what you want to achieve in life. Be very specific and detailed about the things you want to accomplish during your time here on earth. When I say specific, I mean really specific. Don't just write that you want a nice car; name the make and model of car you want. Don't just write that you want a nice house; note how many square feet the house is, how many bedrooms it has, and where it's located. Don't just write that you'd like to travel the world; name the places you intend to visit, the things you hope to do, and the people you want to go with you. List the experiences you want in life and who you would like to share them with.

The more detailed, the better. Once you can start seeing what you want your future to be, you can determine what you need to do in order to get there. This simple process helped me to identify and remove the distractions preventing me from living the life I desired. I have no doubt it will do the same for you.

CHAPTER 1

BURN THE BOATS!

Achieving greatness starts with the right mindset

Tony Robbins is famous for saying, "If you want to take the island, then burn your boats,"[3] but the concept actually dates back to the Roman Empire over two thousand years ago.

As the story goes, the leader of the Roman Empire, Julius Caesar, set out by boat to conquer England. As the massive Roman naval fleet made its way through the channel, they could see the army of Celts that occupied England lining the Cliffs of Dover. Doubt and fear crept into the minds of the Roman warriors as they looked up at the formidable foe outnumbering them ten to one. Caesar understood that if his soldiers' minds were filled with fear and doubt, the battle was lost before it had even begun. But Caesar was committed to this conquest. Retreat was not an option. He needed to get everyone in his army as committed to the cause as he was. In a bold and unprecedented move, Julius Caesar ordered the burning of the ships.

This gutsy move did two things. First, it sent a powerful message to his army. By setting the boats on fire, he let his troops

3 "Tony Robbins," AZ Quotes, accessed February 25, 2021, https://www.azquotes.com/quote/1464549.

know the only way they were going home was by conquering the Celts and commandeering their ships. Second, and possibly even more impactful, it shifted the fear factor. When the chanting and screaming Celts saw the Romans set their own ships ablaze, they knew they were facing an enemy determined to fight to the death.

History looks favorably on Caesar's move. His outmanned army went on to conquer all of England. The burning of the boats marked a point of no return. That one bold action removed all options and galvanized the Roman army around a singular goal: to win at all costs.

How committed are you to the success of your business? How badly do you want to live your best life? Have you destroyed every possible excuse? Have you made success your only option? Are you mentally committed to never accepting failure or giving up? Have you burned the boats? Leaders must create this mindset for themselves, and they have to transmit it to their team.

In this book I am going to give you the exact framework I've used to help hundreds of businesses increase their revenue, optimize their operations, and enhance the stability of their businesses. If you follow the blueprint I lay out in this book, I have no doubt you will be successful.

But you have to contend with an unfortunate truth. Did you know statistics show that ninety-seven percent of people who get a book never read past the seventeenth page? If you want to make the most of this book, you need to be part of the three percent that reads to the end. If you want your business and your life to change in the ways you dream of, you have to commit. You have to burn the boats.

Even if you start with the best of intentions, you will face challenges in putting these tactics to use in your business. It would be all too easy to close the back cover on this book and let your good ideas go to waste. It would be natural to feel that the daily tasks and responsibilities of your life need to take priority over the work of optimizing your business. It would be tempting to give up when you see your quick results start to fade.

If you want to see this through, to make your business the best it can be, to live the life you're dreaming of, you'll have to make some big changes. In order to regain your time and implement these changes with ease and efficiency, you're going to need some help. You have a vision and you can see the path to achieving the business and life that you desire, but you are also very self-aware. You understand your strengths and weaknesses, and you are ready to take action to ensure your success. So, go out and find the people who can help you achieve your goals. Your willingness to take action and your refusal to turn back are the only things that will push you past the challenges awaiting you along this path.

Take decisive action: that is the key. It is the one thing that has separated my most successful clients from all of the rest and helped them to achieve results ten times better than others. They saw the opportunity, analyzed the data, and understood what it would take to make it happen. They took swift and immediate action. Listen, you could have the greatest idea in the entire world, but if you do not put the necessary action behind it to bring it to fruition, what is that idea really worth?

So, the very first tool you need to implement to optimize your business is a shift in your mindset. I want you to commit. Commit to having an open mind. Commit to making changes. Commit to doing whatever is necessary to build the best version of yourself and your business. Commit to taking action.

It's time to burn the boats.

PERSONAL LIFE APPLICATION

Do it or die.

Just do it.

You only live once (YOLO).

The words are different, but the meaning is the same. Stop making excuses and make things happen. These are all great motivational sayings, but they are just words if you don't put action behind them.

Living the life of your dreams is hard, but it's not impossible. Take a look around at all the successful people throughout history and in present day. If you look at their stories, I am confident you will find someone who has experienced challenges similar to your own. If someone before you overcame the challenges you now face, you can do it too.

Every successful person had to overcome some sort of adversity. Success has no biases. It simply requires two things: unrelenting determination and willingness to put in the necessary work. Regardless of your age, your gender, or your race, as long as you are willing to put in the work and never quit, success will find you. The only thing stopping you from living the life you desire, the life you deserve, are the excuses you have chosen to accept. The question is, are you willing to burn the boats? Are you committed to succeeding in spite of your circumstances?

I wrote this book in the midst of the 2020 COVID-19 pandemic. The country was going through one of history's worst recessions. It was being torn apart by racial, social, and economic divisions. Not only did I write this book in less than sixty days, but I added a new product offering for my business and completely revamped our online presence and marketing strategy at the same time. All this despite the world's chaotic state and being personally diagnosed with thyroid cancer at the young age of forty.

While I could not control any of these circumstances, I could control how I chose to respond to them. Rather than feeling down, getting mad at the situation, or giving up, I decided to burn the boats. I decided to lean into the adversity and push myself to succeed in spite of the circumstances, for two simple reasons: impact and legacy.

When I was diagnosed with thyroid cancer, my doctors told me, "A-Jay, I know you don't want to hear that you have cancer, but the good news is that thyroid cancer is the most treatable form of cancer you can get, so it could have been worse." Statistically speaking, that may be true, but what if they turned out to be wrong? What if that wasn't the case for me? I couldn't just sit back, do nothing, and hope they were right. I couldn't count on having more time. If this was going to be the beginning of the end for me, I wanted to make sure I had an impact on others and left behind a legacy that my children could be proud of.

What are you dreaming of? What excuses are you allowing to stand in your way? Burn the boats, start living your dreams, and may the best of your todays be the worst of your tomorrows.

CHAPTER 2

SAT NAV

Enlist the right guide to help you reach your destination

Sat Nav is short for satellite navigation. In the military, satellite navigation is used when planning a mission in enemy territory, moving convoys, or planning search-and-rescue missions behind enemy lines; having reliable real-time information in these situations is invaluable. Knowing the enemy's position, the locations of their facilities, their strongholds, and the position of their troops creates a huge advantage. Being able to step back and see the big picture from a broad strategic level allows us to make quick and calculated decisions to ensure the success of our mission. When you are talking about a matter of life and death, it is vital that you have real-time information so you can quickly make the right decisions. Every minute counts.

You might not realize it, but your technology solutions hold all of the functions of your business together. You should constantly be looking for ways to utilize technology to create a strategic advantage within your business and your market space. If you aren't, there is a very good chance you're behind your competition. That means a loss in market share, stagnated growth, and eventually a loss in customers.

You cannot optimize your business without taking a big-picture view. You need to have accurate information in real time, and you need expert guidance to help you think strategically about how to meet your goals.

Why Technology Matters

The way we do business has forever changed. The speed at which we do business keeps getting faster and faster. Technology, more specifically the internet, is revolutionizing every aspect of the way we do business. Now that we are more connected across the globe, your customers have access to more information and options. They can search the internet and find competitors that aren't even in the same country, let alone the same city or state.

If you aren't sold on maximizing the use of technology within your business, then I have two things you should consider. First, do you think that technology will become easier or harder to embrace in the future? Second, how long do you think you can keep up with your competition if they're maximizing the use of technology within their organization and you aren't?

Let's be honest. Your customers don't even need to search for the product or solution because your competition is probably already bombarding them with advertisements. Have you ever had a conversation about something with a family member, friend, or colleague and suddenly you start seeing ads for those things the next time you're online? This happened to me recently. Last week I was talking to my teenage son about an anime cartoon he was interested in. Next thing I know, I'm seeing ads for anime cartoons popping up everywhere I go online. Someone is always listening.

It doesn't matter what industry you're in, this increase in competition has caused margins to become thinner, which makes each opportunity you go after that much more important. As a business leader, you need to make sure your decisions and investments are strategic and are propelling you toward your desired outcome. Making the wrong move could have a devastating impact on your bottom line and your business. Let me share a story with you about a guy named John (true story, fake name) to illustrate what I mean.

A Costly Mistake

A few years back I was called in to consult with a trucking company. It was the first time I had ever worked with the trucking and logistics industry, so I had no idea what to expect. The owner, John, came out to greet me when I arrived. He told me about his company, stressing that he valued the use of technology. He walked me into his back office, explaining that he didn't hold back any expense when it came to keeping his business on the cutting edge of technology. I thought, *Why is this guy going so over the top trying to impress me?* Curious to see just how much he was exaggerating, I asked John to show me where his equipment was. I assumed his equipment was going to be a mess of dusty cabling.

John guided me to a huge metal door on the west side of the building secured with a numeric lock. I must admit I was impressed by that. I'd expected to find his network equipment sitting on a forgotten shelf in some back corner. He punched in his code, pushed open the door, and stepped out of the way like he was a magician doing the big reveal.

To my surprise, I saw a room full of expensive, high-end equipment. For a brief moment, I forgot where I was and felt like a kid in a candy store. I was amazed as I walked through the room and saw everything he had. After about five minutes, my initial shock started to wear off, and I began asking John some questions. My shock quickly turned into concern and frustration.

I learned that John had invested nearly $250K in equipment that he did not need. He had wired his entire office with a commercial-grade battery backup solution designed to power his entire facility for days should the power go out. It was a great solution. He even had everything color coded and properly labeled. The problem was, John had been there for five years and the power had only gone out twice. Both outages only lasted for a short period of time. Not only that, but John's business was able to function without power because they could easily transition to paper ledgers that could be entered into the system once the power was restored. In the back-left corner of the room, he had a temperature-controlled cabinet! These cabinets are designed to keep your servers and networking equipment at a constant temperature to avoid overheating. It was bolted to the floor and had a lock on the front to prevent theft. Unfortunately, he only had two pieces of equipment loaded in it and the room itself was already well ventilated. As for security, you already know the room was secured by a big metal door with a numeric key lock. When I asked John why he had invested in all of this phenomenal equipment, he simply said, "Because my IT guy told me I needed it."

While making an unnecessary $250K investment is definitely an extreme example, this story highlights four important issues:

1. The IT guy was asked to do something he wasn't qualified to do. There was no doubt John's IT guy was passionate about new technology and understood how to make it all work. The solution he'd put in place was awesome. I was literally in shock when I first saw it. Every piece of this top-of-the-line equipment was properly configured and clearly labeled. It was a really nice setup. But the guy wasn't able to discern between new solutions that would be "nice to have" and

those that were necessary to help John's business grow.

2. John didn't know what questions he needed to ask. He understood that he didn't know enough about technology to manage it himself, so he'd wisely hired an IT guy. However, he should have been asking how each of the IT guy's recommendations were going to help him grow his business. Since his IT guy clearly liked to stay current on the latest trends, he should have asked him what new technology would make them more efficient.
3. John needed someone to help him determine what technology solutions were going to best support his business: someone who understood his business, where he wanted to take it, and what options were the best solution to fit his needs. What he needed was the guidance of a Chief Information Officer (CIO). A CIO would have told John that while the IT guy recommended some great solutions, they were not going to help him reach his goal of opening three more locations over the next five years. That $250K would have been better spent moving his infrastructure into an environment that gave him better mobility and flexibility. Then, when the opportunity for expansion presented itself, he would not be restricted by the confines of his current network deployment.

4. Finally, John didn't have a strategic plan for how his business was going to use technology. Every business leader needs a plan for when and how they will research and deploy new technology solutions within their business. If you're going to upgrade your computers, it should be in the plan. If you're thinking about implementing a new software solution, it should be in the plan. If you're thinking about purchasing a new piece of equipment that will increase productivity, it should be in the plan. This plan is the resource that guides your technology expenses. It's what I call the "Technology Roadmap." Every technology project should be prioritized and planned out by quarter for the entire year in your Technology Roadmap. This allows you to accurately budget for any large expenditures and plan projects around your slower times of the year. Your Technology Roadmap should govern all of your technology projects and should be reviewed on a quarterly basis. This way, you won't find yourself making unnecessary technology purchases throughout the year, you're less likely to have a new technology rollout impact operations during peak season, and you'll stick to your annual technology budget. More on this in the next chapter.

Getting Started

If you're thinking that you're just as bad off as John was, don't lose hope. It took some time, but we were able to get John squared away. He now has complete control over his technology budget and has a plan for addressing technology projects in the future. But getting to that point required John to make some changes. He had to completely change the way he approached the use of technology in his business. You might need to make changes too. For you, this might mean making a major shift in the way your business functions. It might require you to adopt a different mindset. Whatever that change may be, there are three concepts you'll need to wrap your head around if you hope to make this needed transition.

1. **Adjust your perspective on technology.** IT can no longer be viewed as an expense or cost center for your business. You have to view it as a strategic tool that will propel your business forward and possibly create new product offerings for you.

2. **Remove the barriers.** You need to stop thinking of each function of your business as if it's operating independently. Instead, think of technology as being the maestro that harmoniously pulls all of the pieces together. Through the use of next-generation solutions like platform as a service and cloud, you have the ability to create

end-to-end automated solutions. Your options are only limited by your imagination and your budget.

3. **Create a future-proof framework.** While a future-proof system does not literally exist, the concept is important. The way you compete with big businesses in this digital age is by being more nimble and agile then they can be. Therefore, your technology solutions must be able to support you in those efforts.

Finding a Guide

If you're going to successfully shift your mindset about the role of technology in your business, you'll likely need the help of an expert. Here's where finding the right guide comes in. You wouldn't plan a mission into enemy territory without Sat Nav. You wouldn't climb Mt. Everest without a guide. The same applies here: you should seek out someone who is familiar with the terrain and has navigated others through the same process. Someone like the CIO that I previously mentioned.

Before you rush out and start hitting the job boards in search of a CIO to hire, ***STOP***. Most business owners do not need to hire a CIO. I know I just told you that you needed one, so let me explain. The average salary for a CIO in the United States is $175K per year. If you have less than a hundred employees, chances are you don't require the full-time services of a CIO. You need the guidance of a CIO to help you align your technology solutions with the strategic goals of your business. It just doesn't make good business sense to invest in personnel that will only serve you in a limited capacity. Instead of hiring a full-time CIO, what you need is a part-time CIO or virtual CIO (vCIO). A vCIO will provide you with the guidance you need at a fraction of the cost.

A vCIO should be available to:

1. Align your technology to the strategic goals of your business.
2. Vet technology vendors to ensure they can deliver the solutions they promise.
3. Manage your technology vendors to ensure the level of service is consistent.

As you saw with John's story, purchasing the wrong solution can be costly. Choosing the wrong guide can be an equally costly mistake. Not all vCIOs are created equal. Remember, you're looking for a guide who understands your goals and can recommend the best route to get you there. Here are some traits you should look for as you begin your search for the right person/firm to work with.

1. **Business leader:** In order to deliver value and drive revenue, this person should not only understand business strategy but should also be able

to shape it and execute against it to achieve the goals of your business.

2. **Effective communicator:** Communication is possibly the most important "soft" skill this person must possess. They will be working with leaders in all functional areas of your business. They must be able to clearly articulate the objective of their strategy in order to get complete buy-in.
3. **Team leader:** Great leaders do more than just push people to get the job done. They motivate and inspire others to strive to achieve more than they believed they could accomplish themselves.
4. **Creative and innovative:** This person must be at the forefront of our rapidly evolving technology advancements in order to recognize opportunities. They must also be willing to take risks and introduce new ideas and ways of doing things.

The margin for error is getting thinner every day. A CIO or vCIO will help you create a roadmap that will guide how your business uses technology. They will help to ensure that your investments make economic sense and that each solution supports the strategic goals of your business.

PERSONAL LIFE APPLICATION

Every great accomplishment started with an idea that was converted into a plan. You've heard the saying, "People don't plan to fail, they fail to plan." If you want to achieve greatness in your life, you do not need a plan—you need to create a roadmap for your life. A plan is too rigid. It's human nature to get stuck and struggle to figure out what the next best move is when something doesn't go according to plan. A roadmap, on the other hand, is full of options because life is full of twists, turns, and detours. As long as you can see where you are and where you want to be, you can use the map to create a path to get there. You should also review your map on a consistent basis. This will not only ensure that you are still on course, but will also allow you to adjust the course if you find that your desired destination has changed.

CHAPTER 3

TRACER ROUNDS

Illuminate your target with the help of a budget

I will never forget the first time I went through night fire training in the Army. It was a calm, pitch-black July night in Kentucky. The air was thick and muggy. Sounds of wildlife filled the air as we made our way to the training grounds. Apart from the fact that I was out there to do training, it was a rather peaceful moment.

My squad began setting up the Browning .50 caliber machine gun as the drill sergeant gave us instructions. As we worked to set up the eighty-four-pound machine gun, I could see the excitement on my brothers' faces. It felt like we'd been preparing for this moment all day. We loaded up our M16 rifle magazines and the fifty cal's ammo belt with bullets that had little red tips, one red-tipped bullet for every five normal bullets. I could not tell you how many rounds we loaded that day, but it wasn't fun.

Once the drill sergeant had completed his instructions, the training began. From the position we'd taken up behind a bunker on top of a ridge, we anxiously looked out into the dark night, trying to fix our eyes on the targets we'd been told to engage. I remember staring out there, straining my eyes for what seemed like forever until BANG! Soaring high up in the air was a projectile of fire that illuminated the entire area on its slow descent. The moment after that was like something from an old war movie. I could hear nothing but the sound of rapid gunfire and constant pinging as bullets from the weapons of our fifteen-man squad struck the metal targets. The smell of gun smoke filled the air as the bullets, which resembled Jumping Jacks fireworks, danced through the darkness.

Once I saw it in action, everything the drill sergeants had taught us earlier that day became clear. The red-tip bullets that we had meticulously loaded into our magazines and ammo belts were called tracer rounds. When you're shooting in the dark with regular ammunition, you can't see where your bullets are landing. The tracer rounds light up when you fire them. This allows you to see exactly where the bullet is going. Without the tracer rounds, you don't know if you're even close to hitting your target, which decreases your chances of eliminating the immediate threat while wasting ammo and resources that could be used in other parts of the mission.

As a business leader, it is your job to establish objectives and goals to support the growth of your business. That includes having a systematic way to evaluate and utilize technology solutions that will help you reach your goals faster. However, without an effective and reliable structure and budget for expanding your technology solutions, you won't have a clear path forward and you won't be able to determine how well you're allocating your resources. It's like trying to hit a target in the dark without anything to light your way. You may eventually be fortunate enough to hit your goal a few times, but at what cost? How much longer will it take you? How many resources will you waste in the process?

An Expensive Shot in the Dark

A few years back, I was brought in to help a nonprofit organization that was expanding. They had two locations and were about to add a third. They were working with a third-party IT vendor, but they were not confident the vendor could continue to support them as they grew. Curious as to why they felt that way, I began asking my usual questions.

Me: How is their response time?

Client: Pretty good. If we have an issue, they're usually pretty quick to respond.

Me: Are you experiencing the same issues constantly popping up?

Client: No, they're great at solving problems for us.

Me: Are you experiencing performance issues?

Client: We have some slow machines here and there, but for the most part everything seems to work.

At this point, I was really confused. I didn't expect to hear such good reviews about this company since I'd been brought in to find a replacement. So I asked the executive director, "If everything is going so well, why do you want to replace them?"

Client: Don't get me wrong, we like them. It's just that in the past few years with them, we've consistently spent over our IT budget. We won't be able to afford the overages as we expand into our third location.

Me: How much over budget are we talking? Were these overages due to projects or just random expenses?

Client: Last year we went over by almost $20K. We didn't have any major projects taking place. It was just the random replacement of machines that died.

Once I heard that, I knew exactly what their problems were. For starters, they didn't have a documented plan for how the organization was going to manage technology, the Technology Roadmap you read about in chapter 2. If you recall, the Technology Roadmap should contain every technology project that needs to be addressed in order to execute the business's strategic goals. These projects are then prioritized and scheduled by quarter. This document should be reviewed every quarter to track progress. This is also the time to review the plan to ensure the outlined direction still makes sense for the business.

The second problem was that they didn't know how to set a technology budget. Unforeseen incidents happen, but they shouldn't cause you to go over by almost 35% of your total budget. After further discussion, the nonprofit told me their technology budget was nothing more than a loosely calculated guess. The finance officer simply took what they'd spent the previous year and added 5%. While that's better than having no budget at all, it's far from accurate.

In order to increase your bottom line, you need to know exactly what you spend your money on. I'm going to share with you the simple formula I give my clients to calculate their annual technology budget. Then I'll give you an idea of how this budget could be broken up.

Calculating your annual technology budget is simple. Businesses should budget 3–7% of their annual revenue on technology expenses. If you are in an industry that is less technology dependent (e.g., construction, landscaping, dog groomer, etc.), you will be closer to the 3% end of that range. If your business leans heavily on technology (e.g., banking, medical, manufacturing, etc.), you should be closer to 7%. Calculating the annual budget is pretty straightforward and simple. Allocating that budget is a little more complicated and depends on how you strategically use technology.

Your technology budget should be broken up into three categories: Management, Security, and Research and Development (R&D). Let me break each of these down so you have a better idea of what I mean and how you can apply them to your business.

1. **Management.** This part of your budget is what you use to keep your current systems running smoothly. This will include upgrading aging equipment and renewing software licenses and service agreements (e.g., internet, phones,

IT vendor, etc.). You'll also use these funds to account for technology failures and disruptive events. (BONUS TIP: All of your hardware should be on a replacement lifecycle. If you want to avoid unexpected hardware costs, set your refresh cycle to coincide with the end of the warranty. For example, if you don't purchase extended warranties, your computers will be on a three-year refresh cycle.) Without a doubt, this will be the bulk of your budget (75–90% depending on your business).

2. **Security.** Security has been a growing budget item over the years. It used to be that the only technology risk you had to worry about was a system crash or natural disaster. But now, cybercrime has grown to be a 1.5-trillion-dollar industry on the backs of small businesses. To combat this, you need to do an annual risk assessment and penetration test to determine your vulnerabilities. From there, you prioritize these vulnerabilities based on the likelihood of exposure and the severity of the impact it would have on your business. At that point, you try to address the vulnerabilities that pose the greatest risk to your business. This line item comes down to how well you want to sleep at night, but I typically see businesses allocating 10–20% of their technology budget on security.

3. **R&D.** If you're serious about utilizing technology to create a strategic advantage in the market space for your business, you'd better allocate some money to researching and testing new solutions. Every new idea or product initiative requires research and testing. This line item will cover those expenses (0–5%).

Let me put it all together so you can see what this would look like.

For example:

Business revenue = $5,000,000

Technology budget = $350,000 (7% of revenue)

Management = $262,500 (75% of technology budget)

Security = $70,000 (20% of technology budget)

R&D = $17,500 (5% of technology budget)

You should implement this formula into your business immediately if you don't currently have a way to manage your technology expenses. I encourage you to experiment with the percentages to see what makes the most sense for your business. How you allocate the budget is completely up to you, though I would caution you not to stray too far outside the ranges listed. Otherwise, you might find yourself in a situation where you don't have enough money budgeted to meet your needs.

Once you've established your budget, refer to it when you create your Technology Roadmap. This will help you to plan your technology expenses quarter by quarter for the entire year. At the end of each quarter, you should evaluate whether the technology projects you had planned cost more or less than the amount you allocated. Reviewing your budget along with your Technology Roadmap at the end of each quarter will not only prevent your technology costs from getting out of hand, but will also show you whether you're setting accurate budgets. If you find that you consistently go over or under the budget, reevaluate the annual budget and the allocations that you have set.

PERSONAL LIFE APPLICATION

If you find it hard to stick to a budget, chances are you probably have a hard time achieving your personal goals (e.g., lose weight, learn a new skill, etc.). The problem is not you or your ability to set goals. The problem is you've probably never been given a framework for achieving those goals. So here is a simple framework you can use to start achieving your goals so you can have the life you desire.

1. Get a journal and write out all of your goals. It doesn't matter how big or small the goal may be. If it's something you want to accomplish in

your life, write it down. If you've been following instructions, you'll already have a list started. (Remember the list of things I had you create in the first *Personal Life Application* section of this book?)

2. On a separate page in your journal, write down why accomplishing these goals is important to you. It could be because you want to be able to support your parents or provide your family with a better life. It could simply be because it will make you happy. Whatever the reason, write it in your journal.
3. Write down what changes you will need to make in order to achieve your goals. This may require you to learn new skills, meet new people, or stop hanging out with people who are not supporting you. Be honest with yourself about these changes and write them down.
4. Write down what will happen if you don't make the changes necessary to achieve your goals. This is the time for you to be brutally honest with yourself. For example, obesity runs in my family, so one of my past goals was to be healthy so I can see my two boys and two girls grow up. If I hadn't made the changes needed for me to live a healthy lifestyle, there was a chance I could become obese, suffer a heart attack, and die before I could walk my daughters down the

aisle. It may sound extreme, but you need to go to the extreme so you can understand the consequences of not making changes. I know this may be tough, but holding back is only going to hurt you in the long run.

5. Read everything you've written twice a day (once when you wake up and once when you go to bed). This will serve as a constant reminder of what you stand to lose and what you stand to gain. Reading your list every day will keep your goals top of mind and keep you motivated to push through times when you're tempted to give up.

CHAPTER 4

THE TRIPLE S METHOD

Scale your business with one simple method

The United States Army has roughly 480,000 active duty service members. While they perform different jobs, all 480,000 of them went through the same Army Basic Combat Training (BCT). For many, Army BCT is the most demanding and challenging thing they will ever experience in their lives. It is a systematic program designed to transition you, in just ten weeks, from an unfocused, undisciplined individual to a focused, disciplined, mission-driven member of a highly effective unit.

I vividly remember my time going through BCT. By design, the first three weeks, known as the Red Phase, is the hardest part of the transition. The purpose of this phase is to break bad habits and push everyone to come together and start functioning as a team. It's also designed to weed out people who are unable to adapt to military life. The second phase, known as the White Phase, is a little easier. During this phase you learn the basic skills of being a soldier. In the final phase, known as the Blue Phase, you take everything you've learned and start applying it in the field.

The Army has created a proven system that consistently produces the same results. It doesn't matter what your background or experience is before you enter BCT. By the time you complete your training, you're a soldier in the United States Army. Being able to take thousands of people through a ten-week process that completely transforms them is the epitome of efficiency and scale.

The more efficient your business is, the faster you will be able to deliver your product or service. The faster you are able to deliver your product or service, the more time you have to take on additional clients. In today's fast-paced buyers' market, the name of the game is speed. If you cannot quickly respond with what your buyer wants, at the moment they want it, they will move on and find someone else who can. Regardless of the industry you're in, there is just too much competition out there. You must equip your sales team with the tools they need to meet that demand. Still, this isn't enough; you have to take it one step further. Every job function throughout your business must have access to accurate, relevant, and timely data.

If you're in manufacturing, it's vital that you know the number of potential sales that could be closed and where each deal is in the process. Being able to forecast sales allows you to acquire the necessary materials to meet the demand without having excess product sitting on the shelves.

If you have a service-based business, you need to know the number of clients one rep can support without sacrificing your standards. When you're forecasting new client acquisition, you'll know when you need to start hiring new reps to support your growth.

The key to increasing efficiency is having many good data points that paint an accurate picture of how information flows throughout your organization. Your goal should be to understand who needs to have access to what information so you can figure out how to get them that information as quickly as possible. Having this level of visibility within your business will allow you to make quick and confident decisions that will get you closer to your desired outcome.

The Triple S method, which stands for Simple Scalable Systems, is the process I use to help businesses discover and resolve their inefficiencies by:

1. Analyzing the current structure and processes.
2. Identifying inefficiencies and areas for improvement.
3. Implementing a technology solution.

The strategy is designed to help you identify the areas within your business where the strategic use of technology can streamline your processes and reduce errors. When it comes to increasing your bottom line, you only have three options: make more sales, increase your rates, or save money by increasing your efficiency or reducing your overhead (i.e., do more with less). When done correctly, the Triple S method can help you do all three at the same time.

Seeing the Triple S Method in Action

A colleague of mine is the third-generation owner of a heating and cooling company. He reached the point in his business where he didn't need to be involved in all of the day-to-day activities in order for the business to function. At the time, he had one location on the south side of town and twelve service trucks dispatched every day. He wanted to open a second location on the north side of town so he could expand his service area. He knew what he wanted to accomplish, but he was having a hard time mapping out how to make it happen, so he asked me to take a look at his strategy and help him come up with an actionable game plan.

The first thing I had to do was evaluate the operational workflow of his business in order to *analyze* how information passed through the business and understand what they currently had in place. Much to my surprise, and disappointment, they didn't have much technology and automation in place. In fact, the vast majority of their operational workflow still relied on manual processes. On the back office wall was this huge piece of plexiglass with a grid that had names written in black grease pencil. This is what they used to manage the dispatching of their trucks. Each morning the office manager would pencil in the names and addresses of that day's scheduled work. As jobs were completed, they would draw a line through that name and then dispatch that available service tech to the next name on the list.

All around the office were filing cabinets and mounds of boxes full of paper files. I had never seen such an organized yet chaotic office in my life. Every job they had completed over the past thirty years could be found on a paper work order filed in one of the hundreds of boxes that filled their 1,500-square-foot office space.

Needless to say, there were a number of inefficiencies that needed to be addressed before they could expand. As I *identified* these for my colleague, he was surprised; he didn't realize how much time and money was being wasted because of simple mistakes made in their manual processes. The office manager would order the wrong parts because she couldn't read the service tech's handwriting. Fuel and time were wasted when a service tech was dispatched across town to handle a job when they already had another tech in the same general area.

It took some time, but I finally convinced him to invest in technology that would streamline his business. There were two things we had to immediately address. The first was their dispatching. By *implementing* an online dispatching system with tracking, the office manager was able to easily see where the trucks were in relation to all of their jobs so she could quickly dispatch the techs to the jobs closest to them. The system also gave them in-depth statistics so they could see how long service techs were spending on each job. They were then able to use this information to determine if they had a tech that needed additional training signaled by long service times on the same type of work orders. The second was an inventory management system. This allowed them to easily track the parts they had in inventory and helped to reduce errors when it came to ordering custom parts.

After it was all said and done, my colleague was able to dramatically increase the efficiency of his operations. Over the course of about seven months, he was able to expand his operations and service more clients without having to drastically increase his workforce. As business increased, he was even able to gradually increase his rates. This is the power of implementing scalable processes and solutions to increase the efficiency of your business.

Making the Triple S Method Work for You

So how do you start using the Triple S method for your business? Start your analysis by detailing your customer's journey. You need to have a clear picture of the entire process from prospecting to pitching to onboarding and servicing your clients. You need to know what information you're collecting, how you're collecting it, and where it's being stored. To do this, you need to evaluate every aspect of your business:

- Talk with your sales team to understand the process they use to convert a cold prospect to a warm lead, and then take that warm lead and convert them into a closed deal.

- Work with your operations team to understand what the handoff from the sales team looks like and what the onboarding process is for a new client once the deal is signed.
- Get with your finance team to evaluate the handoff from the sales and operations teams to ensure that clients are billed properly for the agreed-upon product/service.
- Finally, talk to your account managers to understand the process for resolving issues, upselling existing clients on additional services, and asking your clients for referrals.

As you're talking to your staff to understand how information flows through all of these areas of your business, here are some questions you should be asking:

- What are the tasks that you complete on a daily basis?
- How long does it take you to complete these tasks?
- What information do you gather that needs to be given to other functional areas within the business?
- How does that information get from your department to the next?
- Are there any processes in the current workflow that we could automate in order to increase efficiency and reduce errors?

In smaller organizations, one person may be responsible for multiple parts of the process, which is fine. The Triple S method is designed to flush out points of friction regardless of how big your organization is. A point of friction in your workflow would be anything that slows the client acquisition process down. You're weeding out inefficiencies—any outdated process that makes the client jump through unnecessary hoops or makes tasks harder on your staff. Look for things like manual processes or steps that require duplicate entry. The goal of this exercise is to identify all points of friction within your current workflow. There is a very good chance that you can automate these processes with the right technology. The more processes you can automate throughout the customer's journey, the easier it will be for you to scale your business and obtain predictable results.

If we look back at my colleague's situation, for example, he had a huge blind spot when it came to his inventory. He had roughly $15K worth of inventory sitting on his shelves because of all the wrong custom orders placed over the years. By implementing a digital inventory management system, the service techs were able to see what parts they had in inventory while they were onsite with the customer. If the part they needed was not in stock, they could submit the order request for approval on the spot. The service tech was now able to set clear expectations to the customer. In turn, this led to a better overall experience for the customer, which led to an increase in business, which allowed them to increase their service rates.

As a business leader, it is your responsibility to make sure that everything is working as efficiently as possible. Don't get stuck in a level of comfort or complacency. Just because something has always been done a certain way doesn't mean it can't be improved. Your management should review and question your processes on a consistent basis. If they have an idea for a better way of getting the work done, then that solution should at least be considered. If you simply apply the Triple S method within your business, you'll find areas that you could optimize that will result in an increase in performance and revenue.

Maximizing the efficiency of your business is critical to successfully growing your business at scale. Scaling your business is about being able to consistently produce predictable results in larger quantities. Just like Army BCT, you need a systematic process you can plug anyone into and know they will produce the desired results if they follow the process. Strategically deploying technology solutions will allow you to create these predictable results. However, with the endless number of technology options available, it is critical that you find the solution that best suits your needs. We'll discuss this in greater detail in the next chapter.

PERSONAL LIFE APPLICATION

The Triple S method is about streamlining tasks and communications. Every one of us has access to the tools we need to streamline structures in our personal lives as well as our business systems. Unfortunately, most of us don't realize it. Thanks to the power of your smartphone, streamlining processes in your personal life has never been easier. For example, utilizing a shared calendar is without a doubt the easiest way to reduce friction between you and your family and friends. You can create calendars allowing them to see your availability and add items to your schedule. Instead of waiting for a free moment to talk about scheduling an event, your spouse can check your availability and add something to the calendar.

This is one basic example of how you can bring more efficiency to your life by using tools that you already have. Think outside the box and look for ways to use technology to automate tasks or streamline communications in your world.

CHAPTER 5

STEP INTO DOMINANCE

Evaluate technology solutions to elevate your business

Have you ever wondered why the United States military is considered the strongest military force in the world? Our military is the best on the planet because we have the best technology. Our tactical naval, air, and ground equipment is second to none, and information regarding their development is a prized target for cybercriminals.

During my time in service, one of my jobs was to provide communication support to a tanker unit. While I thought our field communication equipment was cutting edge, it was nothing compared to the technology in the tanks. I was fascinated by the level of precision and accuracy of their weapons guidance systems. A tank commander once bragged about how accurate his guidance system was, saying that his main gun could fire a forty-one-pound armor-piercing sabot round and hit the tip of my nose from two miles away while on the move. That level of distance and accuracy blew my mind. Especially since I was struggling to consistently hit the three-hundred-meter target with a single rifle shot while stationary.

If you desire to dominate your market space, your business needs to be able to do two things—maximize profitability and run at peak efficiency. In order to achieve these two things, you need to have the right tools in place.

In the previous chapter, I explained how to use the Triple S method to find processes within your business that you could automate through technology. But with countless technology solutions in the market all claiming to satisfy your needs, how do you determine which solution makes the most sense for your business?

Based on my experience, most business leaders don't have a process to evaluate potential solutions. I have seen business leaders use nonsensical metrics for choosing solutions: the friendliness of the sales rep, the lowest price, the outcome of a coin toss, the colors of the company logo.

With new solutions hitting the market every day, each one promising to solve your unique problem, it's easy to see how business leaders get entangled in decision-making. We're talking about multiple applications within the business that may or may not communicate with each other. Each one boasts an impressive list of marketed features you probably don't need and will likely never use. Without careful planning or consideration, technology can quickly become a liability to your business rather than an asset. Let me explain.

The Hidden Cost of Technology

A residential/commercial contractor was experiencing some great success and growth. Six months prior to meeting me, he had hired six new employees. To manage the additional workload of these new hires, he'd been forced to upgrade his server hardware. He'd purchased a new server to increase his storage capacity and performance. While the new hardware was a dramatic improvement on what he had before, there was a major flaw. The owner didn't realize that the server he purchased had a fifteen-user limit. At the time he brought me in, he had thirteen employees and was in the process of hiring four more employees. That would give him two more people on his staff than his equipment's capacity. This was the equipment he had purchased only six months prior. When I asked him why he'd purchased this specific hardware, he said it was a good deal at the time and the tech told him it would meet his needs.

Based on the information given, the tech had told the truth; he'd had no idea the contractor's business was going to grow that quickly. The contractor had no idea that a user limit even existed. *Because he did not have a clear process to evaluate technology solutions*, the contractor was forced to make another large capital investment and endure the sunk cost of the investment he had made six months ago.

The STEP Method

The STEP method is the framework I use to help business leaders evaluate their technology options. STEP stands for Simple Technology Evaluation Process (STEP). This framework, which consists of three questions and ten criteria, will help you invest in a solution that meets your current and future needs.

First, determine what you need the solution to do. Create a wish list of requirements that you would like for the solution to have. You will not find a solution that checks all the boxes, so you will need to prioritize your wish list.

Organize your features by three categories:

- Must have
    - These are features that are deal breakers. If the solution doesn't have all of these features, it's not an option.
- Would like to have
    - These are your optional features. It is not a deal breaker if a few of them are not there, but the solution needs to have most of these items.
- Would be cool
    - These are your luxury features. You don't need any of these features, but it would be a bonus if a solution had them.

Three Questions and Ten Criteria

Once you have your requirements categorized, you're ready to start the evaluation process.

I. How will this solution solve my immediate needs?

How you answer this question dictates whether the solution makes it to the second question in the evaluation process. If it cannot immediately solve a business problem, there is no reason for consideration. To further determine the solution's worthiness, measure it against these three criteria:

1. **Features:** Does the solution provide features that meet your requirements? If not, can it be customized to meet them?
    - Do not instantly dismiss a solution if it doesn't tick all of the boxes. There may be other advantages, such as flexibility or overall value, that offset the things it does not do.
2. **User-friendliness:** Will this solution be easy for your employees to learn, implement, and adopt?
    - Avoid the trap of assuming you can determine this on your own. While you may think you know how your employees will respond, it's best to bring them in to help you evaluate this piece. During your evaluation, don't forget to

look at what impact the solution will have in other areas of your business. The solution may be great for the user capturing the data on the front end, but is it easy for the user managing the data on the back end?

3. **Security:** Are you able to secure your data both internally and externally?
    - Internal security – Are you able to set permissions that restrict access based on user description or job function? This will ensure that staff members only have access to the things that are pertinent to their job. The solution should also have audit logs that allow you to track when changes to data are made and by whom.
    - External security – You want to be sure your data in this system will be protected from the outside world. Look for features like single sign-on, two-factor authentication, and the ability to enforce strong password requirements. If you are storing data on the provider's cloud, be sure to get documentation that explains how they protect their servers both physically and digitally.

II. How will this solution meet my future needs?

This question is equally as important as the first question. The solution needs to be able to grow with you. It does you no good if your business outgrows the solution in two years' time. Use these four criteria to determine if the solution can scale as your business grows:

1. **Flexibility:** Determining the flexibility of a solution can be tricky. Here are some of the things you should consider:
   - How configurable are the automation features?
   - Can you customize the analytics?
   - Can the data sets be customized to meet changing analytic requirements?
2. **Compatibility:** Does the solution seamlessly connect to tools you're already using (e.g., email, office products, calendars, accounting, etc.)?
   - The more tools the solution will communicate with the better. Even if you never need to use the integration, the fact that it already exists gives you options in the future.
3. **Innovation:** How many new feature updates have been released in the last year, and how helpful were those new features?
   - The best solution providers invest heavily

in R&D to ensure their products stay ahead of the curve when it comes to features, user-friendliness, security, flexibility, and interoperability.

4. **Ecosystem:** Is this a new solution? Does the producer have a strong list of industry partners it works with?
   - A good way to determine the stability of the producer is by evaluating the number of customers and partners it has. If it is a brand-new solution that has not been proven yet, you are likely to experience issues and may have trouble getting support.

III. Does the solution fit my budget?

To accurately judge the cost of a solution, you must look at the total cost of ownership. Total cost of ownership breaks down into the following components:

1. **Setup costs:** How much will it cost to get the solution up and running?
   - Setup costs include consulting services required to customize and train your employees as well as the time required for your employees to help design, test, and adopt the tool.
2. **License costs:** What is the annual subscription cost (if there is one)?
   - Make sure you understand exactly what you

are and are not paying for. Many software providers automatically include features and benefits you may not need into their initial quote. Do your homework to make sure you're not paying for a feature you will never use.

3. **Maintenance costs:** What will it cost you in time, resources, and money to support this solution?
    - A good rule of thumb: the more customized your solution, the more it will cost you to maintain it.

Using the Decision Matrix

Now that you have the framework, put it all together to fairly evaluate each solution. Start by creating a decision matrix. Score each prospective solution from 1–10 on each of these criteria, where one means the solution does not meet your requirements and ten means it exceeds your expectations.

Here is an example on the following page of what your decision matrix could look like:

The Decision Matrix

Characteristic	Weight	Tool 1	Tool 2	Tool 3
Features	20%			
User-friendliness	15%			
Security	20%			
Flexibility	15%			
Compatibility	10%			
Innovation	10%			
Ecosystem	10%			
Weighted Score				
Cost of ownership over 5 years				

Feel free to adjust the weights so they reflect the criteria that are most important to your business. Once you determine the weighted percentage for each criteria, multiply the weight by the score you assigned and add them together to get an overall weighted score for each solution. Then take the annual subscription cost and the projected annual maintenance costs, add them together, multiply that number by five, and add in the purchase price and setup costs for each solution. This will give you the total cost of ownership over a five-year period.

Applying the STEP method will give you an objective and balanced way to evaluate future technology investments. It will also help to ensure that you invest in a solution that will scale as your business grows.

PERSONAL LIFE APPLICATION

If you struggle with making complicated or big decisions, you can use the STEP framework as an objective way to evaluate your options. You don't need to create a decision matrix every time, although you certainly can. In most cases, you can insert your situation into the three main questions and then make note of your responses. An example of this could be:

Do I pay $1,000 to take a personal development course?

1. How will taking this personal development course help me now?
2. How will taking this personal development course help me in the future?
3. Does the $1,000 investment in myself fit into my budget?

Simply running your decisions through those three questions will help you slow down the decision-making process in your mind and keep you from making impulse decisions that you may later regret.

CHAPTER 6

TURNING ON THE LIGHT BULB

Develop a framework that brings it all together

When I was in the Army, I was a communication specialist. In the beginning of my career, it seemed as though my days were filled with trivial tasks. If I wasn't doing maintenance on our equipment, I was running troubleshooting drills or practicing setting up field communication systems with the rest of my platoon. I thought setting up field communication systems just to turn around and tear them down by the end of the day was a complete waste of time. But that changed after I experienced my first field training exercise with the entire battalion.

I had never experienced anything like it before. It was like watching a construction crew build a large building. There were roughly a thousand soldiers out there, each performing the specific job they were trained to do. Once I observed everything in motion, I could see how each function was dependent on another function. For instance, the commanders relied on my platoon to establish communications for the Tactical Operations Center (TOC), but we couldn't set that up until we knew where and what the TOC layout was going to be. Understanding how everything worked together and how important my role was to the bigger mission helped me to maintain my discipline and stick to my training.

By now you know I recommend working with a vCIO to create a Technology Roadmap, using a formula to set appropriate technology budgets, and implementing the Triple S method to identify areas within your business where technology solutions could increase efficiency. Now it's time to pull all of those pieces together as part of your operations. Use this five-step process as an operational framework for managing technology in your business:

1. Establish your technology budget and use the Triple S method to identify the areas of inefficiency within your business.

2. Prioritize these issues in order of importance. Remember that the Triple S method was not designed to be completed alone. Make sure to get employee feedback about the solutions you're prioritizing. Figure out what would make their life easier while still making an impact on the overall efficiency of your business. Not only will this ensure that you're focusing on the correct issues, but it will also increase employee loyalty. Each issue you prioritize now becomes a future technology project.

3. Create your Technology Roadmap. You now have everything you need to create your Technology Roadmap. Look at your list of projects and determine in which quarter of the year you can initiate each project based on your budget and your

peak busy season. PAY ATTENTION HERE. It's not realistic to attempt to tackle every project in a calendar year, so don't put that pressure on yourself.

4. Use the STEP method to identify and evaluate solutions to address the first project on your list, consulting your budget and Technology Roadmap as needed. Once that project has been completed, review and document how the project went. You'll want to make note of things such as:
   - Was the project completed on time?
   - Was the project completed on budget?
   - Has the solution resolved the issue as expected?
   - Did implementing this solution create other issues we hadn't previously identified?
   - What lessons did we learn through the process?
5. Review your Technology Roadmap to determine if the next project on the list still makes sense for the business. If it does, proceed to using the STEP method to identify and evaluate solutions for that project. If it doesn't, go back to the Triple S method to determine if there is a more pressing issue that needs to be addressed and proceed from there.

I hope you can see now that this process is ongoing;

it never stops. What I'm outlining for you is an operational framework for not only managing technology, but also for maintaining business efficiency. As your business grows and technology evolves, this framework will put you in a position to identify those gaps and areas of inefficiency so you can quickly adapt and make the necessary changes.

PERSONAL LIFE APPLICATION

Living the life of your dreams and being the best version of yourself is a process, not a destination. It's a process because everything is constantly changing. The person you are as you read this today is dramatically different from the person you were five years ago. And the person you will be five years from now will be dramatically different from the person you are today. Life will throw you curveballs. Your wants and desires will change. It's all part of the process called life. Do you have an operational framework for life that can grow with you? If not, consider adapting the framework from this chapter to fit your personal life. These tools are designed to help you navigate this process so you never get to a point where you feel stuck in life. Change is an important part of growth. So embrace the change, seek the change, and never stop burning the boats!

CHAPTER 7

BASIC TRAINING

Create effective training to ensure a qualified and loyal staff

I woke from a deep sleep to the loud ringing of clashing metal. I sat up and opened my eyes to see a large figure standing over my bed beating a metal trashcan lid with a tire iron. Everyone was rushing out the door, so I quickly gathered myself and followed suit. Out in the cold dark night, I expected the yelling to stop, but it didn't. Instead, I found more people barking orders at me. It was 2 a.m., and I was standing outside in shorts, a T-shirt, and a pair of socks. This was my first night of my first week in basic training.

Army basic training consists of three phases, strategically designed to accomplish four things:

- *Break your old habits*
- *Bring you into their culture*
- *Teach you their way of doing things*
- *Teach you how to work as a team*

Regardless of how you may feel about it, it is hard to argue with the results. The Army's basic training program does a phenomenal job at transforming new recruits into soldiers and getting them to adapt to military life. Their training program is successful because participants only have one choice. Adapt to the Army's way of doing things or go home.

As a business leader, I'm sure you replicate the Army's goals for every new hire training. (BONUS TIP: if your current new hire training is not set up to accomplish these four things, I would encourage you to rework your training program.) Unfortunately, most of us don't have the luxury of letting people go within the first sixty days of employment. This is why we have hiring managers to ensure we're looking at quality candidates.

For example, let's say you're looking at hiring an accountant to manage your business finances. You have two candidates. The first is trained and has a degree and years of documented experience. The other was not trained, does not have a degree, but is familiar with the tax code after handling their personal taxes for years. Which one are you going to hire to manage your business finances? The first candidate is the clear and obvious choice. But what if the second candidate was sixty percent cheaper and required no "up-front" cost? You still wouldn't hire them, would you? Why not? Because they lack the skills and experience needed to validate that they can perform the job you need them to do.

A business leader wouldn't intentionally hire an

unqualified candidate. Yet so many business leaders continue to employ underqualified workers. Most of the time, employees become underqualified because of changes in technology or because a better process was discovered. But that is still no excuse for not training people. The two biggest excuses business leaders have given me for not providing training are:

"We are too busy to take time off for training right now."

"We just don't have the money to pay for training."

I understand where these sentiments come from. Training requires a business to invest its two biggest resources: time and money. You have to pay for someone to lead the training. You have to pay the employee while they're attending the training. In addition, unbillable hours and lost productivity must be accounted for. Over the years, I've also learned that business leaders don't want to pay for their employees to receive continuing education or training because they had a bad experience. They hired a company to provide training and it was done poorly, or the material they covered just didn't help. I get it; investing in training that fails to deliver comes at a high cost. I don't blame any business leader for not wanting to risk that again.

However, you also pay a price when you choose not to train your employees. Here are five factors you may not be considering that will help highlight the importance of training.

1. **Mistakes cost time and money.** We all know that improperly trained employees make mistakes. Every mistake costs the business time and money. You then lose more time correcting the mistake or you damage your reputation by delivering an inadequate product to the client.
2. **Untrained employees are inefficient.** Poorly trained employees take longer to complete tasks simply because they're not aware of the best way to complete the task. In some cases, this hampers productivity by as much as 40%. Imagine what you could do if your business was able to increase productivity by 40%.
3. **Untrained employees produce lower-quality work.** A person that has been trained on how to create presentations in PowerPoint or to generate complex reports in Excel will always produce a better presentation than an untrained person. Not only will they create higher-quality work, but they will complete the task faster.
4. **Untrained employees equal unhappy employees, which equals high turnover.** Imagine that untrained employee we've been talking about: making mistakes, taking longer to complete tasks, and consistently underachieving. If that was your experience eight hours a day, five days a week, fifty-two weeks a year, how

happy and motivated would you be? Eventually, you would get tired of it and look for something better.

5. **Untrained employees lead to lost customers.** The compounding effect of the mistakes, poor quality, and inefficiency will eventually impact your customers. It may manifest in the delivery of a bad product or the missing of a deadline. Regardless of how it manifests, the likely result will be the loss of a customer.

Properly training staff leads to employees with new skills that can improve production, cut time spent in creation of your product (or service), reduce production costs, reduce mistakes, build confidence in your staff, and create a better working environment. An investment in developing your employees' skill set is an investment in your company.

I hope you can see the value in training your employees and the cost of not training them. Now let me show you how you can use my **BASIC** Training method to easily develop a cost-effective continuous training program within your business.

Be inclusive

Ask your staff for their ideas about the type of training they would like to receive. You may not be able to see what areas your staff need to work on most. By getting them involved, you'll be able to provide the tools they need to improve their skills. Highly skilled employees will always yield better results for your business. Even if you can't incorporate every suggestion into the training program, including employees in the process has a powerful effect. It makes your employees feel seen and lets them know their opinions matter. Your staff will work ten times harder for you when they feel like they're important and valued by the team.

Alternative training methods

Technology is constantly evolving, presenting us with new ways to learn new skills. You should regularly search for the latest training tools, such as relevant apps and online courses. Free courses are available across many academic and professional websites. Remember that training doesn't always have to be technical. Consider bringing in guest speakers from businesses and colleges; they can discuss their expertise and provide a different type of learning opportunity to your staff.

Set a budget

If you're operating with a lean budget, start where you can and find alternative training methods to maximize the use of every dollar. Regardless of how much you start with, you must include training as a line item in your company budget. This will help your team see that you care about their advancement so much that you have made it a justifiable and measurable cost.

Identify your leaders

Every company has a person (or group of people) who just wants to learn new things. I bet you can see that person in your mind right now. See if they would be interested in helping you build out a training program. Having a champion who researches, plans, and implements most—if not all—training activities can help keep the cost down. Giving them a leadership role also helps shape their management skills in the process.

Cross-train

The most overlooked resources are those that already exist within your business. Cross-training and shadowing are great ways to help your employees learn new skills. This approach also allows you to see if some of your employees perform better in other areas. Not to mention, it's a great alternative to investing in an expensive formal training program.

Never forget that your employees are the lifeblood of your business. It will always be easier and cheaper for you to train the talent you already have than to attract new talent. Remember that your training program should constantly be evolving. As your company grows, so do the needs of your employees. If you continue to emphasize the importance of training, you'll develop a highly efficient staff and create a culture they'll never want to leave.

PERSONAL LIFE APPLICATION

Never stop learning. The internet has given us access to an abundance of free knowledge. The only thing you need to invest is your time. Take the lid off of your imagination. You can literally learn anything your heart desires if you set your mind to it and put in the work. Google and YouTube are your portals to becoming whatever you want to be. The only thing stopping you is you. Don't let limiting beliefs stop you from taking action to pursue your dreams. These beliefs will only lead you down the path of regret.

If you do have funds to invest in your personal growth, you might be surprised at the affordable costs of online courses. With the rise of influencers and experts on social media and online tutoring platforms, you can find a wide range of educational content that is sure to fit any budget. Find a program that feels right for you but, most importantly, just get started. The best investment you will ever make is in yourself. It's an investment that will always pay dividends.

CHAPTER 8
STOP

Enforce policies and procedures that get results

"Cease fire!" The command instantly put an end to the sound of rapid gunfire echoing throughout the range. Two drill sergeants made a beeline down the row, one from each end. They didn't look happy, and they were heading in my direction. My mind raced, trying to figure out what I had done wrong. They looked like they were going to tear me a new one. With a hard, cold stare, the drill sergeant walked past me to meet up with the other drill sergeant three stations down to my right. Without a word, he took the weapon away from the private that was occupying that lane, pulled the magazine, and cleared the chamber.

Then he yelled, "Private ... what is the standard procedure for handling a weapon that misfires?"

The private responded, "Drill Sergeant ... the standard procedure for handling a weapon that misfires is SPORTS ... Slap, Pull, Observe, Release, Tap, Shoot ... Drill Sergeant!"

"And where should the barrel of your weapon be pointed while executing this procedure?" the drill sergeant barked back.

"Down range, Drill Sergeant!" the private replied.

While we were practicing firing from the prone position at the range, the private's weapon had misfired. When he went to perform SPORTS, he had inadvertently turned his weapon down the row, putting us all at risk. The drill sergeants explained that the two main reasons for having standard operating procedures were for speed/efficiency and safety.

You need to have policies and operating procedures in place for your business. They ensure that everyone is performing a task the same way in order to get consistent, predictable results.

Most of your policies and procedures are probably outlined in your employee handbook (BONUS TIP: if they are not, they definitely should be). I'm going to give you the benefit of the doubt and say that you do have your policies and procedures outlined in your employee handbook. I'm even going to give you credit for reviewing your policies and procedures with your HR person on a consistent basis. But let me ask you, how strictly do you enforce your policies and procedures?

I've seen countless organizations that don't uniformly enforce their policies and procedures. It's become so common that I created the acronym STOP to serve as a reminder for my clients. STOP stands for Stick To Operating Policies. On the surface it may not seem like a big deal, letting some of your staff color outside of the lines if it better suits their workflow. But let me show you why sticking to the operating plan that you laid out is so important.

The Creative Disaster

I was doing some work with a creative design agency that had the coolest office I've ever been in. I still remember the first time I visited. I walked in through the front double doors into this spacious reception area. The receptionist sat behind a beautiful wooden desk with a black stone top. Behind the receptionist were four fifty-inch digital displays that rotated through some of the projects the company had worked on.

Past the reception area, I got to the fun stuff. There was this huge open space filled with oversized beanbag chairs, lounge chairs, and a couple of small tables. There was a foosball table along the left wall, and at the back of the room I saw three arcade games and two pinball machines. On the right was another set of double doors that led to the creative space. When I walked through the doors, I felt my mood just magically change. The room was filled with earth tones and wood accents. There were no florescent lights in that area, just the soft glow of desk lamps and a few artistic floor lamps. The smell of incense or potpourri lingered in the air and calming music was playing. Being in the space made my logical and analytical brain want to be creative.

Their office was designed to be laid back and promote collaboration and creativity. I think it's one of the reasons they attracted such talented people; they did some amazing work. But it was this same laid-back approach that led to a mistake that cost them thousands in weeks of lost productivity, thousands more in remediation costs, and ultimately a couple of clients.

The company had policies in place that outlined the use of personal devices and third-party applications. Essentially, employees were allowed to use a personal device for business work as long as the IT team confirmed it had proper security measures in place. They were absolutely not allowed to use third-party applications that were not provided by the company to work on business files. Unfortunately, the company was not very good at holding people accountable to these policies. A user wanted to test a new third-party application, but the work network kept blocking access to the program. So the employee used their phone as a hotspot, connected their work device to their personal hotspot, and used that connection to access this new program they wanted to test. The user didn't realize they were hitting a corrupted link and unknowingly infected their machine with ransomware. After connecting the machine back to the work network, the virus made its way to the shared folder where it encrypted their ten years' worth of data. It took them three weeks and close to $20k to get everything back up and running again. During that time, they also lost three clients because they weren't able to meet project deadlines. Had this business been enforcing the policies they'd already outlined, this incident would have been avoided.

Putting Policies in Place

As a business leader, there are two things you need to take away from this story. The first is to ensure that you have good policies established. Here are some of the policies that you need to have in place:

- **Bring Your Own Device:** This policy outlines the use of personal devices within your business.
- **Bring Your Own Access:** This policy outlines the use of personal hotspots within your business.
- **Application Usage:** This policy outlines the use of authorized and unauthorized applications within your business.
- **Device Usage:** This policy outlines how your employees are meant to use the technology (e.g., laptop, desktop, printer, scanner, etc.) that is given to them. This policy is where you restrict or limit the use of business resources for personal use.
- **Internet Usage:** This policy outlines the appropriate use of the internet service provided by the business. This is where you would prohibit the use of streaming services like Netflix, Hulu, and ESPN. I have seen work productivity in some environments drastically decrease during March Madness or major golf tournaments.

Every business should have these five policies in place as a bare minimum. You should also have policies in place that help promote security, but I'll go into more detail about security in the upcoming chapters.

The second thing you need to take away from this story is that you must hold people accountable. Once you have the policies in place, remember to STOP. Talk to your HR manager to create an appropriate disciplinary process for enforcing these policies. If you're not enforcing the policies you have in place, you will have a hard time holding someone accountable when their violation leads to a catastrophic incident.

Most inner office issues arise from a simple misunderstanding. By having policies in place that harbor open lines of communication, you reduce the chance of conflict; the key to avoiding and resolving conflict is communication. An added bonus is stronger bonds among your team members. When you know that the person beside you is working to achieve the same thing you are and you have a way to address misunderstandings, you develop trust and accountability within your team. Just like in the military, you'll accomplish more when your team is working together as a cohesive unit.

PERSONAL LIFE APPLICATION

You don't have to be as strict in your personal life as you are in business. In the business world, you can create clear expectations and dictate how people are required to function within your business environment. If they choose not to comply, there are clear consequences. Your personal life, on the other hand, is full of different types of relationships. You have family, friends, acquaintances, and strangers, none of whom are under your control.

So what do you do when you can't control the actions of others? You create a set of standards for yourself, a group of "policies" that you hold yourself accountable to. These policies govern how you act toward others as well as how you allow others to interact with you. Just because you can't control how another person behaves toward you, doesn't mean you have to accept it. This is a value I work diligently to instill in my children. No matter what the relationship is, you must have clear boundaries set for yourself that govern what you're unwilling to do and behavior you're unwilling to tolerate from others. Do not bend or waver once you've established these boundaries for yourself. If you do, you may eventually find yourself dealing with a catastrophic situation that will cost you more than you're willing to pay.

What are your personal policies? Where do you draw the line? What boundaries do you refuse to budge on?

CHAPTER 9

ERA

Evaluate risk to optimize your business for evolving markets

It was a hot and humid summer day in July, and I was attached to an antidrug unit working in conjunction with the state police. Our mission was to traverse the rough and rugged terrain of the foothills of eastern Kentucky to locate, confiscate, and destroy marijuana crops. Flying above us was a two-man spotting crew that would radio us the coordinates of the next plot for us to seize. We had gone as far as we could go in the vehicles and would have to travel the last twenty-five hundred feet or so on foot to our next destination.

*My squad and I could find no clear path up the hill to the coordinates we were given. I have to admit I was not paying attention as I was walking. Suddenly, my squad leader shoved me to the ground. My squad leader was a six foot one, two hundred and thirty pound special forces guy who had a bunch of patches on his chest. To put it bluntly, he was a badass motherf****r. I was shocked as I looked up at him from the ground. Then I was pissed off.*

I jumped up to see what in the hell was going on. When I

got to my feet, he pointed out the danger that lay before me. There were two traps set where I'd been about to step. The first was a bear trap. The second was a trip wire that would have released a board full of nails at my head. This revelation shook me to my core. I had been seconds away from danger, and I hadn't had the slightest clue. Sensing my discomfort, my squad leader pulled me aside, looked me straight in the eye, and told me something I will never forget. "Orr, you have to be aware of everything around you. The greatest threat to the success of any mission you'll ever be on is the threat that you didn't know existed. Not knowing could be the difference between everyone coming home safely and your brothers coming home in a body bag."

In the business world, we oftentimes categorize threats as risks. While the risks we face in business may not carry the possibility of loss of life, an unknown risk could cause the death of your business. If you think about it, every business that has ever failed did so because it didn't know how to respond to a risk that materialized.

As a business leader, it's your job to manage risk in your business. You must understand that your risks are constantly changing. There are environmental risks, digital risks, market risks, political risks, and countless others that may be unique to your industry. It's important that you Evaluate Risk Annually (ERA). Let me show you what could happen if you decide that proper risk management is not a priority for your business.

The Shortsighted Boss

I was fresh out of the military and had just started my career, working for a small technology firm that provided support to other small businesses. At the time, our method of support was a break-fix model: the client would call us any time they had an issue, and we would send someone over to fix the problem. The client would either prepay for a block of hours they could use, or they would pay by the hour when they needed us. The benefit of purchasing a block of hours was the hourly rate was discounted, and they received priority service whenever they needed us.

To my mind, this entire process screamed of inefficiency. I could never plan out a week's worth of work because any client that had purchased a block of time could call in with an emergency. That would completely throw off the schedule for the day because their service call took priority over what was already scheduled. So I started researching different ways of providing our services. At that time, technology firms on the West Coast were offering a proactive service: they charged the client a monthly fee in exchange for proactively monitoring their network and resolving little issues on a monthly basis. This helped to prevent the little things compounding to create big issues that ultimately resulted in downtime for the client.

I loved the idea and knew my clients would love it too. Wanting to understand how they were doing it, I reached out to a couple of providers in California. I told them I was part of a small firm in Ohio and was interested in offering a service like theirs in my market and would appreciate any advice they were willing to share. They were excited and honored to share their knowledge with me—I hadn't expected that. Each of them opened up their playbook and showed me everything they had tried that worked. They also told me what hadn't worked and why it had failed. I took detailed notes and corresponded with them via email for weeks to get their input as I put my plan together.

I created a proposal that outlined how I would present it to the market, why I thought it was a better solution for our clients, and how it would allow the company to increase their revenue. I scheduled time to meet with the owner (remember, this was a small company) and show him what I'd learned. I remember the sense of pride I felt as I gave my pitch. I was sure it was going to be a slam dunk. I got to the end of my presentation and asked him what he thought. He might as well have laughed in my face and called me stupid. He said that I was not being paid to think of business strategy, I was being paid to keep his clients happy. I argued that we were in the technology industry, the fastest changing industry on the planet, and we would be in trouble if we didn't change as the industry changed. He said he knew his clients, that they would never go for something like this, and that I should stick to what I'd been hired to do.

I quit the job that day in my mind. I was confident the program I had laid out was going to be the way that all technology support firms would do business. Little did I know this was the early stages of today's managed services providers. I left the firm a few months after that conversation. Two years later the owner had to shut his doors because he refused to accept the risk of the market changing. By the time he realized his error, it was too late, and he could not respond fast enough.

Developing the skills to effectively manage risk is important to the success of your business. Every great business leader, past or present, has been phenomenal at taking and managing risk. The good news is that risk management is a skill you can learn and develop.

Evaluating Risk in Your Business

There are four steps to risk management: Identify Risk, Evaluate Risk, Develop a Response Plan, and Execute. Let us take a closer look at each one.

1. **Identify Risk.** The very first thing you have to do is identify your risks. A risk is anything that could have an effect, positive or negative, on your desired outcome. I typically start by identifying every vulnerability in a project or situation that would cause it to fail. This step is the most important when it comes to managing risk. It doesn't matter how ridiculous the potential risk sounds. If it can impact your desired outcome,

write it down. The goal at this stage is to identify every risk regardless of the probability.

2. **Evaluate Risk.** Risk evaluation allows you to separate and prioritize the risks you have identified. This way you can focus your attention on the risks that will have the greatest impact. You should evaluate each risk based on the impact it would have and the probability that it will actually happen.

   Go through the list of risks you identified and assign each one a value. Rate the impact and probability on a scale from 1–5, with 1 being "very unlikely" on the probability scale and "negligible" on the impact scale and 5 being "almost certain" on the probability scale and "catastrophic" on the impact scale. Then multiply the probability and impact scores to create a priority grade for each risk of Low (1–3), Medium (4–6), High (8–12), or Extreme (15+). Here is an example of what your risk assessment matrix could look like:

Risk Assessment Matrix

		Impact				
		Negligible 1	Minor 2	Moderate 3	Major 4	Catastrophic 5
Probability	Almost Certain 5	Medium 5	High 10	Extreme 15	Extreme 20	Extreme 25
	Likely 4	Medium 4	High 8	High 12	Extreme 16	Extreme 20
	Possible 3	Low 3	Medium 6	High 9	High 12	Extreme 15
	Unlikely 2	Low 2	Medium 4	Medium 6	High 8	High 10
	Very Unlikely 1	Low 1	Low 2	Low 3	Medium 4	Medium 5

Let's look at how my previous employer would fit into the matrix. If he had admitted the possibility that the industry could change, that would have been a 3 on the probability scale. If he looked at the situation objectively, he would have seen that failing to adapt would have had a catastrophic impact on his business. That would have given him a 5 on the impact scale. Multiplying the two together, he would have gotten a score of 15, making it an extreme priority that he should have addressed.

3. **Develop a Response Plan.** Congratulations, you have evaluated the level of impact and probability of each risk and have assigned each a grade. Now is the time to determine how you will respond to these risks. When it comes to minimizing the impact a risk can have, the faster you respond, the better off you will be. You should have a documented response plan for every risk regardless of the grade you've given it. You need to know what you're going to do before a risk occurs so you can immediately execute a response. There are four ways you can respond to risks:
    - **Accept:** Accept that this risk may occur. These are typically the risks you gave a Low evaluation to. No need to do anything with them because, if they do happen, the impact is not severe or can be easily overcome. When the benefits of success outweigh the cost of failure, accept the risk and move on.
    - **Avoid:** If the risk could have a catastrophic

impact (Extreme), the best solution may be to avoid it all together. Determine everything you can do to prevent this risk from occurring. If you can't find a viable solution, you may not want to take on the task or start the project at all. This is the safest option, but it should also be your last option.

- **Transfer:** Look for ways to transfer the risk to someone else. If you're not skilled at doing a necessary task, hire an expert. By doing so you have transferred the risk of failure over to them (e.g., hiring an accounting firm to manage your business finances).
- **Reduce:** When the impact of the risk is too much to accept and you cannot avoid it or transfer it, find a way to minimize the impact of that risk. A good way to reduce the risk in a project is to bring on partners. Hopefully, you can bring on partners with more experience than you have and thereby reduce your exposure to the risk. They may be able to find a way to help you exploit the risk.
- ****Bonus** Exploit:** This applies more to projects. In some cases, you can actually exploit the risk to bring about a benefit. Put another way, find the opportunity within the risk. For example, say you sell software and a company wants to use your software but their

old equipment can't run your program. You can provide them with new hardware or make margin by bringing in another company to provide that service.

4. **Execute.** This is where the rubber meets the road. Proceed with business as usual, constantly looking for any event that will trigger one of your documented risk responses. Move toward your desired outcome while monitoring for risks. If one occurs, execute the response you created for that risk.
5. *****BONUS*** Document and Learn.** During and after the execution of your risk response plan, you should take note of the outcome. Did your response work? What could you have done better? In hindsight, do you see things you could have avoided or opportunities you could have exploited? By reflecting on the experience, you'll be better equipped to handle risks in the future. The better you get at managing risk, the more comfortable you'll be taking on more risks.

Risk does not exist in a vacuum. As your business progresses and grows, so do your risks. Therefore, you should constantly be looking at how these changes add or eliminate risk. At a bare minimum, you need to evaluate your risk annually. Over time and through many lessons learned, your risk management skills will improve. The more you go through this process, the easier it gets and the faster you become at recognizing and evaluating risk.

Remember that, even though risk is always present, and the consequences can be destructive, you can create a plan. There are ways to guard against risks, to prevent them, and to minimize their effect if and when they occur.

PERSONAL LIFE APPLICATION

It's been said that the best things in life are on the other side of fear. Whether you subscribe to that theory or not, I'm sure you can agree that you need to take some risks if you're going to live the life of your dreams. This process for evaluating risk is universal and will help you to evaluate any risk in life. Having a planned response to a threat before it happens allows you to execute your response faster and minimize the potential impact. It also helps you to be aware of your situation. Because you have thought things through, you will have a better idea of what the threat looks like so you can recognize it before it gets to you.

Don't let the fear of risk hold you back from pursuing your dreams. Brainstorm about the things that could prevent you from achieving the life that you desire. Make a list of possibilities. Plug them into the risk assessment matrix, create a response plan, and charge after your dreams.

CHAPTER 10

MYTH BUSTERS

Recognize real threats and protect your business from cybercrime

In the early 2000s, there was a show on television called MythBusters. In each episode, the two hosts took a common myth and used science to either validate the myth or debunk it. I've always been fascinated with physics and chemistry, so the show was an instant hit for me. I wasn't the only one; this popular show ran for eighteen seasons. The hosts had an uncanny way of taking complex things and making them understandable and logical.

By now you know how to identify the risks that pose a threat to the success of your business. If cybersecurity is not an item on your list, you need to pay very close attention to what I'm about to say. In our digital world, where everything about us and our businesses exists in a digital format, cybercrime and data loss are potentially the biggest threats to the success of your business. In 2019, cybercrime was a 1.5 trillion-dollar industry with forty-six percent of attacks targeting small businesses. If that wasn't bad enough, studies show that sixty percent of small businesses that suffer a data breach file for bankruptcy within six months.

This phenomenon results from misconceptions about cybersecurity shared by many business leaders. The biggest one? My IT guy has me taken care of. Listen, cybersecurity is not just the concern of IT professionals and large corporations. We all rely on technology to perform tasks that are essential to the operation of our business. At the end of the day, you must remember one thing: hackers target people, not computers. Computers are just the tools they use to get to us. Cybercrime is on the rise and small and mid-sized businesses continue to be targeted.

I want to start the conversation about defense by addressing some of the most common myths around cybersecurity. My goal is to help you understand the severity of the risk you face and how it's constantly changing. This way you can start to address your vulnerabilities and have a response plan in place before a cyberattack occurs.

Cybersecurity Myths

Consider the following top six cybersecurity myths:

Myth #1: My IT guy/team/company has it taken care of.

If you don't have a dedicated team of security engineers constantly watching over your business, you are vulnerable. Simply put, most businesses do not have a skilled expert on staff to adequately handle security. And that is not a knock against any IT person. Almost everything runs on the data network in your office. In the past it was only your computers, but now you have phone systems, camera/alarm systems, printers, and smart devices all using your network to access the internet. Most IT departments are spread too thin because there is just too much to stay on top of. Most IT departments are responsible for making sure the PCs, servers, printers, mobile devices, and phones work; on top of that, they have to worry about the vulnerabilities created through digital marketing and social media platforms. Take it from a guy who is thoroughly entrenched in the IT world: I get how challenging it is.

Finding time to keep up with your basic skills is challenging enough without adding in advanced cybersecurity skills. Figuring out how to maximize the use of technology within your organization is a challenge. Understanding what you need to be protecting in the future is a challenge. And to top it all off, you are still expected to keep the systems running and put out every fire that pops up.

The reality is that most IT departments are so busy maintaining the system, they don't have time to think about improving security, let alone actually do it. Criminals understand this challenge and they use it to their advantage. It's the reason they're always three steps ahead of the current security model. They don't have to worry about maintaining a network, making sure everything works, or resolving issues that come up.

Nope, they simply write as many pieces of malicious code as they can, fire them out there, and wait for the system to let them know they've hit a target.

Myth #2: We would immediately notice if our systems were hacked.

Ten years ago, I would have agreed with this line of thinking. At that time, if a computer was infected, it showed noticeable signs: The computer started running slow. Pop-ups appeared all over the screen. Sometimes it became completely unresponsive. However, cybercrime has now grown into big business for the criminal underworld. Their methods have evolved and if you don't have a way to rapidly detect when something has gone wrong, it's too late. By then you're in trouble: expect to be on the news or find yourself out of business.

It doesn't matter what industry you're in, what you're doing, or what the situation is. When it comes to minimizing damage, the name of the game is speed. Think about it: if you could detect and locate a fire in the forest right when it starts, you'd have a much better chance of stopping it before it grows into a raging wildfire.

The same holds true in data security. Unfortunately, the average data breach in the US goes undetected for 206 days. That's almost seven months of someone having access to your company data without your knowledge. Let me put that in perspective for you.

A regular Major League Baseball season is only 187 days.

It only took NASA 158 days to send a probe to Mars.

Hell, you and I could take a luxury cruise around the world in 106 days.

We could sail around the world twice before the average business detects that their systems have been compromised!

Even if you're twice as good as the average business out there, that still leaves you vulnerable for three and a half months. So, ask yourself, how many records could be compromised in just three and a half months? I think the better question may be, how many records do you have on your system? When you look at it from that perspective, you understand why data breaches are so expensive and why sixty percent of small businesses don't survive them.

Myth #3: We're too small to be a target. I don't have anything worth stealing.

It may be true that your business isn't being targeted, but you're still at risk. The vast majority of cyberattacks are untargeted attacks. Criminals simply write a bunch of programs designed to find vulnerabilities in machines or programs to trick a human into clicking on something. Once their program finds a vulnerability and gets inside a business, it alerts the cybercriminal. That's when the real attack begins. The cybercriminal takes the next seven months, on average, going through all of your stuff looking for anything they can monetize. That could be your files, your financials, your intellectual property, your identity, and even your employees' identities. Can you imagine the lawsuits you would face if your employees' identities were stolen simply because you had a data breach?

That is phase one of their attack. After they've looked through and copied everything valuable, they move to the second phase: asset control. In phase two they use your resources to launch other attacks. This could mean sending emails out to all of your contacts. It could mean using your computing resources to launch larger organized online attacks. I once saw a criminal using space on a company's server to store child pornography.

The bottom line is you are always a target. Even if you don't have something a criminal can directly monetize, which is almost impossible, gaining access to your computer resources helps them in other ways.

Myth #4: We have antivirus and a firewall in place, so we're taken care of.

Before you can understand the error in this way of thinking, you first have to understand what you're up against. A new cyberattack takes place every 3.9 seconds. A new virus is created every four seconds.

This means criminals are attacking quickly, often, and from multiple angles.

The traditional approach to cybersecurity is not fast enough to adapt to the evolving threats we face. Have you ever thought about how the current security model works and noticed that it doesn't make sense? Think about it: our security measures are dependent upon someone finding a vulnerability in a program, operating system, or piece of hardware. In other words, we *wait* for a criminal to find the vulnerability and exploit it. Only once the manufacturer or developer becomes aware of the issue do they develop a security patch to plug the vulnerability without reducing the functionality of their product. Who knows how long that process takes, and in the meantime you're still vulnerable.

Even when the patches are released, they're not instantly applied. They sit there until you or your IT person installs them—assuming that either of you is aware that a security patch even exists. You must then repeat this process for every device and every piece of software in the organization. That's how businesses are currently being protected.

Take that long, drawn-out, manual approach and stack it up against the overwhelming velocity that cybercriminals are attacking with. The sheer volume of attacks we face forces us to constantly be on the defensive, trying to plug holes. This stacks the deck against you. If there is only one message you take away from this section of the book, let it be this: The attacker only has to get it right once for their threat to work. Our defensive measures have to get it right every single time.

If that weren't bad enough, we are defending ourselves with one hand tied behind our backs: the same hardware and software we put in place to defend ourselves is also available to be purchased by cybercriminals. This means they can test and fine-tune their attacks against the same resources we use to protect ourselves.

Myth #5: All of our stuff is in the "cloud," so we don't have to worry about security.

This could not be further from the truth. Sure, the cloud is more secure in the sense that the physical servers are better maintained and protected than in a traditional business. But the cloud provider doesn't ensure the protection of your data. If your files get corrupted, that's on you. The service provider is not responsible in the event that a cybercriminal gets the credentials needed to access your cloud environment.

Unfortunately, this happens all too often because people have a false sense of security when it comes to being in the cloud. Logic should tell you that if a cybercriminal is able to compromise a machine within your organization, then it's only a matter of time before they are able to gather the credentials needed to access your cloud environment.

Myth #6: I'm good because I don't open emails from people I don't know.

Ninety-one percent of all data breaches start with an email. This makes it the number one method criminals use to get inside businesses. It's easy for them to send out massive amounts of email; they only have to get lucky once. All it takes for a criminal to get in the door is one person accidentally clicking on a bad link.

We're not talking about a fake email from a Nigerian prince who wants to send you a million dollars. Those scams are easy to spot. Now end users are being bombarded with messages that look like things they actually want to click on—messages that look like they are coming from vendors, shipping agencies, and even financial institutions. Not to mention that it's quite easy for a criminal to copy (a.k.a. spoof) a person's email and make it look like you are receiving a message from someone you know. Therefore, the idea of only opening emails from people (you think) you know, is not foolproof.

On the surface, some of these myths may seem harmless. But I hope you can see how each of these misconceptions might have a critical impact on your business. Managing digital risk is everyone's business, no matter what industry they're in or what kind of business they're running. Do not let these outdated ideologies give you a false sense of security.

As you begin to formulate a plan to address your

digital vulnerabilities, make sure you're not falling prey to the myths we've discussed in this chapter. Now that you're seeing clearly, you can focus on addressing the two most important things: your technology infrastructure and your people. In the next two chapters, we'll go into more detail about what you need to do to secure each of those areas within your business.

PERSONAL LIFE APPLICATION

Living the life of your dreams is hard. You'll face many challenges, but your biggest opponent is yourself. You'll wonder whether the long days, heartaches, and disappointments are worth the effort. You'll doubt that you're good enough and ask whether this is all a big mistake. Don't believe the voice of self-doubt ringing in your head. I'm here to tell you that you can do it.

The biggest myth you'll need to let go of is the myth of limiting beliefs. I don't care what your background is, how many disadvantages you think you have, or how many times you think you've failed. The only time you ever really "fail" at something is when you give up. The rest are all lies and excuses that we tell ourselves so that we become comfortable with mediocrity.

Back when I was going through BCT, an injury put me in danger of not graduating. The situation wasn't fair and the challenge that lay before me seemed impossible. But my drill sergeant pulled me aside and said something profound that has always stuck with me.

He said, "Orr, I know this shit fucking sucks, but listen to me and listen good. Fucked up shit happens to everyone. There ain't shit you can do about it. That's called life. But what you can control is how you choose to respond in the moment. At the end of the day, I can't tell you what to do. But I will tell you this. This moment, this moment right fucking here, you were made to overcome this exact moment. The question is whether you believe that or not."

To this day, whenever things get hard and I'm ready to give up, I remember what he said. I remind myself: "This moment, this moment right here, I was made to overcome this exact moment." I remember that I was built for this, that this is why I am successful when others are not, and I push through it.

It doesn't matter what you're going through. I know it seems hard and the challenges seem impossible to overcome. But you were built specifically to overcome these challenges. You are stronger and more capable than you could ever imagine. Now get out there, complete the task, and live the life you deserve.

CHAPTER 11

OVERWATCH

Choose the right team to provide digital security

I was once attached to a military antidrug unit that worked in conjunction with the Kentucky State Police. In the foothills of eastern Kentucky, the reconnaissance team had come across one of the largest plots of marijuana crops they'd ever seen. It was estimated that this plot had over three hundred plants in it. They had been watching the area for weeks, hoping to catch the owners of the plants coming back to check on them. But when they failed to detect activity, the reconnaissance team pulled out and my unit was given orders to retrieve and destroy the plants. I don't know what the street value was, but with three hundred mature plants, I could only assume it was a lot.

The plot was located in a small clearing at the top of a ridge. In these types of situations, approaching a high value target with multiple points of access, it was typical for us to perform the mission with two teams. The first team was a ground team responsible for the actual confiscation—I was a part of this team. The second team was an overwatch team that constantly surveyed the area from a helicopter above us. Their job was to help guide our movements on the ground and inform us of any foreseeable threats that we were set to encounter.

The plan was simple—approach the location from the south, secure the crops, exit the area the same way we came, and then proceed to the burn site to destroy the crops. As we were making our way up the hill through the trees and thick brush, we received a call on the radio. "Bravo unit, this is Overwatch. Come in, over." My entire unit froze and shifted their attention to the voice coming through the radio. After my squad leader acknowledged the call, the male voice on the other end responded, "Bravo unit, this is Overwatch… be advised that you have a group of possible hostiles approaching your target destination from the north. Halt while we attempt to engage, over."

The next thing we heard was a loud deep voice coming through a megaphone saying, "This is the Kentucky State Police. You are entering a restricted area. Stop right there … do not advance." By that point, my unit had already taken cover and was surveying the area with weapons drawn. The voice on the megaphone came on again, "I repeat… stop right there… do not advance or we will open fire."

The offenders made the wise choice to heed their warning, and they turned around and fled the area. Overwatch advised us when the area was clear, and it was safe for us to proceed. We completed the mission while Overwatch radioed other Kentucky State Police units to come and apprehend the fleeing suspects.

In tactical situations, your overwatch is there to watch your back, and to provide you with protection and support so that you can complete your mission.

As a business leader, you need to understand that your business is constantly at risk of being a victim of a cyberattack. Cybercriminals continue to launch new and more sophisticated exploits with ransomware, viruses, phishing, and denial-of-service attacks that can spread through your network and computers like wildfire. This puts you in the position of constantly playing defense. If you don't have the right solutions and expertise to proactively detect and respond to these invisible threats, then your data and your business are in jeopardy. Just one error can lead to an event that could have a devastating impact on your business.

The Traditional Approach to Cybersecurity

Unfortunately, if you are relying on basic cybersecurity defensive measures, your business is vulnerable. Period. I know that might be hard for you to believe, but the prevention security model and practices the industry has adopted over the past fifteen to twenty years are fundamentally flawed and no longer work as well as they used to. Let me explain.

First, let's assume every business today has some variation of the basic cybersecurity defensive equipment in place: commercial-grade firewall, intrusion protection software, URL filtering, email filtering, and antivirus. These are the most basic things you should have in place to protect your digital assets from threat. But here is the catch that most business leaders don't realize: The basic defensive equipment you put in place is designed to keep out the things that you or your IT team have configured to keep out. When properly configured, these solutions work together and do a pretty good job at keeping out those *identified threats*. But what about the threats you don't know about? How do you detect and stop those threats?

The threats we face are constantly changing. There are already more devices connected to the internet than there are people in the world. Mobile devices, wearables, thermostats, and security automation sensors are all connected to the internet. In addition, we're starting to see Internet of Things (IoT) devices expand beyond the consumer market. More industrial applications for IoT devices are appearing as manufacturers and logistics companies look to IT-based systems in an effort to reduce costs and increase efficiency.

Unfortunately, cybercriminals have already started using these IoT devices to launch attacks. Back in 2016 we had the Mirai Botnet attacks, which used vulnerable IoT devices to bring down the internet on the East Coast. A casino in Las Vegas got hacked through the use of a fish tank connected to the network. A fish tank. A college was attacked through their smart vending machines and smart lightbulbs. And who can forget about the infamous hack of the Jeep Grand Cherokee back in 2015.

A New Approach to Cybersecurity

The point is that criminals are shifting their attacks away from where we watch and defend. You cannot stand over every employee's shoulder telling them what they should or should not click on. And endpoint protection does not exist for these IoT devices. No one is watching these devices, and the manufacturers don't even know what type of security they should be wrapping around them. This is only going to get worse as we see more devices introduced to the network.

How are you supposed to overcome these deficiencies? How can you protect yourself against an enemy that is constantly evolving?

First, we need to change the way we think about and approach cybersecurity. We approach cybersecurity the same way we approach protecting a building. Your firewall is the digital version of a security gate establishing a perimeter and the remaining filters act as an elaborate home security system with sensors and trip wires set to alert you when something bad happens. The problem is that you deploy these solutions and sit back, thinking you're properly protected. But remember, cybercriminals have proven that they can get around these defensive measures and keep their presence hidden for *seven months*.

So I suggest a different tactic. Think of your new approach to cybersecurity more like the human body. How is it that you can always tell when you're coming down with something before you actually ever get sick? Because you know your body. You know exactly how your body normally feels and can tell when something starts to feel off. So, you take some extra vitamin C, consume more fluids, or do whatever it is that you normally do to avoid getting sick.

This same tactic works with your business network. With the right technology solution in place, you can monitor everything on your network and create a baseline for how everything normally operates. When a machine starts to behave abnormally, you can isolate the device and determine if it has been compromised or if it's simply experiencing a hardware failure. This type of solution will dramatically increase the speed at which you detect threats and your ability to adapt to change. By moving to a behavior-based security model, you'll be able to see and respond to threats in almost real time.

Creating Your Customized Digital Overwatch

I suggest that you hire a dedicated group of individuals with a specific set of skills and equip them with unique tools to create your own digital overwatch solution. This approach will help you address the concerns of IoT devices, new and evolving attack methods, and insider threats. The digital overwatch solution I'm describing is commonly referred to as a

Security Operations Center (SOC), and it functions as the first line of defense for your business.

Let me be completely transparent here—building out an SOC is not an easy task. It's a complex undertaking, but I've created a quick breakdown of the three essentials:

1. **People.** The most important piece to deploying a functional SOC is your people. You need a team of highly trained and certified people who are familiar with analyzing security alerts and assessing vulnerabilities. Since digital risks and threats are constantly evolving, you need creative thinkers and problem solvers who can adapt quickly. Attacks can come in a variety of forms and from multiple angles, so it's important to hire people who can learn on the fly.

   Just to be clear, this is not the type of responsibility you give to someone without experience. I could go out and purchase a long-range precision rifle, but that doesn't make me a combat-ready sniper. This team is going to be responsible for keeping your business safe. Make sure they have security credentials and experience to back it up.

2. **Processes.** In order to have a successful SOC, you need to understand the security controls outlined in regulations such as NIST, PCI, HIPAA, and many others. There are a number of security controls they all have in common, but each

regulation has a few unique controls that need to be met in order to be compliant. Not only do you want to be familiar with what the controls are, but you want to know how to remediate them as well. Being able to rapidly detect a security issue is worthless if you can't properly respond and remediate the issue.

3. **Technology.** The true value of an SOC is the ability to detect a security incident in real time and respond and mitigate it in near real time. In order to accomplish this, your SOC must have two primary pieces of software.

   The first is a Security Incident and Event Management (SIEM) solution that identifies, monitors, records, and analyzes security events in real time. This is the tool that gathers all of the data about your network and allows you to create the baseline I mentioned. When properly configured and managed, a SIEM allows you to detect security incidents in real time. The second is an Incident Management System (IMS). When a security incident does occur, it's vital that the right people get notified as quickly as possible. The IMS manages these alerts so that the appropriate response plan can be executed as quickly as possible.

Two Roadblocks to Implementing Your SOC

Imagine having a security solution in place that could easily adapt to the way your digital risks change as your business grows and threats evolve. How much faster could you execute, knowing that you had a team watching over your steps and engaging the enemy before they had a chance to hurt your business? You have to wonder why every business doesn't have an SOC. There are two limiting factors that contribute to the scarcity.

The first is a global shortage of certified security experts. Security experts are in high demand as there are far more jobs available than there are qualified people to fill those jobs.

The second limiting factor is money. Even if personnel were available, most small businesses can't afford to employ a team of security experts that resembles what I've described. That doesn't even include the cost of the tools I mentioned, which can also be a sizeable investment.

An Alternative to the Traditional SOC

If building your own SOC isn't feasible, I encourage you to find a security provider that offers an SOC as a service. Similar to a vCIO, this vendor should be able to provide the benefits of a private SOC at a fraction of the cost. Here are four things to consider as you evaluate your options for SOC as a service:

1. **Credentials:** You need to know what kind of training, certifications, and skills their security engineers and analysts have. Remember, you're hiring them so you don't have to build your own team.
2. **Availability:** Criminals don't work a nine-to-five shift or take weekends and holidays off. Therefore, your provider should be available 24/7/365.
3. **Frequency:** Monitoring and scanning can vary, especially when it comes to cloud environments and on-premises equipment. Be sure you know if they provide real-time monitoring or if they're only doing weekly, monthly, or quarterly scans.
4. **Location:** Where is their SOC located? Their team could be anywhere in the world. Some regulations require that your services are maintained within the United States.

As business leaders, it's our responsibility to reduce risk and mitigate any threats that could damage the business. Given the rapidly changing landscape, having a digital overwatch in place should be a vital part of your strategy. When deployed correctly, an SOC will help protect your business from the threats you don't see coming.

PERSONAL LIFE APPLICATION

The concept of an overwatch is all about teamwork. You should have skilled people in your life providing "overwatch" for you as you navigate the countless obstacles on your path to success. That could be a mentor that guides you through the decision-making process as you grow your business, or a close friend who is there to lend a hand or an ear when times get tough. No matter the situation, you need good people you can count on to help you deal with the complexities of life.

But this concept is a two-way street. You should be providing overwatch to the people who are close to you. It doesn't have to be anything complicated or formal like a mentorship. It could be as simple as providing someone you care about with a different perspective that could help them deal with the challenges they're facing in their own life.

Having people in your life watching over you to ensure that you stay on the right path is an important ingredient to success that many people don't talk about. If you're tackling life's challenges by yourself without support, you have a huge vulnerability. Life is a game of stringing together a series of correct choices. It's extremely hard to make the correct choice when you only have one perspective of the problem.

So who is providing overwatch for you today?

CHAPTER 12
THE FIRE GUARD

Train your employees to prioritize cybersecurity

During my time in the Army's basic training program, there was a duty rotation called "Fire Guard" that everyone had to go through. It was a one-hour shift during the night where you and your battle buddy stayed awake to keep watch over the barracks. The term "Fire Guard" originated back when the barracks were made of wood and were heated using a wood-burning stove. The Fire Guard watched the stoves to make sure the barracks didn't catch fire. While the threat of fire is less of a concern in modern barracks, the tradition remains as an exercise in discipline. During their shift, the two-man crew patrols the barracks looking for recruits trying to leave, watching out for fires, and completing cleaning duties.

In my third week of training, it was time for the floors to be stripped. It was one o'clock in the morning, and my shift had come to an end. My battle buddy and I woke up the next two soldiers to take over Fire Guard duty. We had already detailed the latrines and taken out the garbage on our shift. It was their job to strip the floors in the hallway. I had already put the scrubbing pad on the floor buffer for them. We pointed in the direction of the cleaning closet, told them everything was set up for them, and went to bed.

I'd just gotten into a good sleep when the drill sergeant came in screaming and ordering everyone to get out of the barracks. Tired, dazed, and confused, we all fell into formation outside. We stood there waiting to see what sort of unusual punishment was about to be cast down upon us. The drill sergeant took his position after ensuring the barracks were clear and gave us the at-rest order. He asked how we were all feeling and if we were okay.

The soldiers who had relieved us on Fire Guard duty had mixed bleach and ammonia to make a stronger cleaning agent in order to clean the floors faster, not knowing the combination was dangerous. They'd created two large batches of lethal chloramine gas.

Everyone was okay, but I share this story to highlight two points. Number one, you don't know what you don't know. These soldiers were operating with good intentions. They had no clue that what they were doing could have such a devastating impact. Number two, ignorance can have dangerous outcomes. If the drill sergeant hadn't come upstairs and recognized the situation, we all would have been exposed to that lethal gas longer, resulting in permanent damage or even death.

When it comes to protecting your business, your people will always be the weakest link in your security strategy. Humans are prone to making mistakes, forgetting things, and falling for sneaky criminal tactics. Here is the scary part that most people never talk about: you can have the most advanced defensive security solution in the world, but it can all be bypassed if a user gets tricked into doing something they shouldn't. This is why training your employees on the latest criminal tactics and keeping security top of mind is so important.

Why Your Employees Need Security Awareness Training

If that doesn't convince you of the importance of having a security awareness training program, then maybe one of these five reasons will.

1. **It reduces the possibility of data breaches.** When people hear about a data breach, they automatically think of computer hardware, like a criminal "hacked" into a computer or network to steal sensitive information. But in reality, the number one thing that leads to a data breach is human error. Every fancy security system businesses use to protect themselves can be bypassed if a user clicks on the wrong link, goes to the wrong website, or falls for a trick and unknowingly hands over their login credentials.

   This is why implementing a security awareness training program is so important. Your employees are the last line of defense for your business. Educating them on the latest tactics that criminals are using helps to form a "human" firewall for your business.

2. **It keeps your organization in compliance.** On the surface, this seems straightforward, but you'd be amazed at how many regulated organizations miss this. If you're in a regulated industry, you're required to have a security awareness training program. It doesn't matter what security regulation you're governed by, they all have a rule or statute that states that security awareness training is a requirement in order to be compliant.

   The tricky part is they don't stipulate what this security awareness training program needs to include. So I often see organizations throwing something together and slapping a security awareness training label on it just to check the box. But this approach misses the spirit of the requirement: to educate your staff on the dangers out there so they don't unknowingly become a victim. If your goal is to simply check off a box, then your staff is no more aware than they were prior to the "training." This means they are still a liability to your business.

3. **You'll save money.** The average cost of a data breach on a small business ranges anywhere from $149K all the way up to multiple millions. I constantly have people asking me if a data breach could really be that costly to their business. Absolutely!

There are a number of things people overlook when they think about the potential cost of a data breach to their business. From my experience, the biggest thing they fail to consider is the risk associated with their digital assets. People think their cost and exposure is limited to *remediating the incident*, but there are other potential repercussions to consider, such as:

- Lost revenue due to theft
- Lost revenue due to your damaged reputation
- Lost revenue due to clients leaving
- Lost revenue due to downtime
- Cost of legal fees to fight lawsuits
- Loss of market advantage due to theft of intellectual property
- Loss of personal assets due to identity theft

And this is just a short list of things that every business has in common. Once you start to factor in those costs, you can see how data breaches become extremely expensive to a business.

4. **You'll increase employee confidence and reduce stress.** What are the consequences for an employee who makes a mistake that leads to the business being compromised? Do they get written up? Do they get fired?

   If you're not training your employees on how to

avoid the kinds of mistakes that could get them fired, you could be putting unnecessary stress on them, which can result in decreased performance or poor judgment. This strain can ultimately lead to the very thing you want to avoid. Provide your staff with the training they need to identify these threats so they can be confident in the job they're doing.

5. **Cybercrime isn't going away.** Cybercrime is a trillion-dollar business. That's right, I said *business*. Let me put it in perspective for you. Criminals make more money from cybercrime than they do from the sale of all illegal drugs combined. Plus, cybercrime carries less risk. So if you could make more money and have less exposure, would you ever stop? No. In fact, you'd likely increase your efforts to find more creative ways to capitalize on the opportunity. Don't leave your employees vulnerable to this ongoing threat.

How to Set up Security Awareness Training

Hopefully, I've convinced you that providing security awareness training is a good idea for your business. How do you get started? What information should be covered? Here are six things you should consider for maximum impact as you implement a successful security training program.

1. **Start with your policies.** A good security awareness training program starts with clear policies and procedures for your staff to follow. Your security awareness training program should serve two purposes: 1) to educate your staff on the digital threats that exist and the tactics that criminals are using, and 2) to reinforce the reasons behind the specific policies and procedures in place in your business.

   If you don't have security policies and procedures laid out for your business, here are some basic policies you need to consider adopting.

   - **Acceptable Use Policy (AUP):** The AUP defines what an employee is and is not allowed to do with the technology resources provided by the company, to include internet access.
   - **Password Policy:** The Password Policy defines the password standards for your company. Include guidelines for password length, complexity, how often passwords need to be changed, and how long you must wait before repeating a password.
   - **Cybersecurity Policy:** A Cybersecurity Policy is an umbrella policy that can address multiple topics. It can provide guidelines for how to handle confidential data, how to protect personal/company devices, safe email practices, and procedures for remote access.

- **Bring Your Own Device Policy:** A Bring Your Own Device Policy defines what types of devices your employees are allowed to use to access company resources, what resources they are allowed to access, and what security measures must be in place before they can use that device.

2. **Remember that it's all about the people.** As I mentioned before, your cybersecurity awareness training program should not be about simply checking a box. Your program should consist of fun and engaging topics that are relevant to your people. Not only does your content need to be informative, but it must also be easy for your people to understand and execute. The goal should be to inspire dialogue in a safe and open environment. Over time, this type of culture will encourage more conversations and help your staff take their own digital risk seriously.

3. **Cover current and relevant topics.** Having informative material is worthless if it's not relevant. Every four seconds, a new virus is created. Your cybersecurity awareness training program needs to be updated consistently. Your people need to be educated about the latest tactics criminals are creating to attack businesses. If you fail to regularly update your training content, you might as well skip the training altogether.

4. **Include real-world simulations.** If you want your cybersecurity awareness training program to truly be effective, include a test. Not only should you assess your people on the material they're learning, but you should also perform random simulated tests. In the real world, your people will be required to make decisions that determine whether the organization gets breached or not. Conducting random simulated testing will condition your staff so they have a better chance of making the right decision the next time they're faced with a challenge.

5. **Create a method for accurate reporting.** You need to be able to see if your cybersecurity awareness training program is closing the knowledge gap. Detailed reporting allows you to see how effective your training program is and will show you which members of your staff are having the biggest challenges. Then you can provide them with remedial training to help them better protect themselves. Good reporting will also give you the insight needed to optimize future trainings. By looking at past results, you'll be able to see what is working well and what can be improved upon.

   Be consistent. How often should you be doing security awareness training? In regulated industries, the laws vary on how frequently you should conduct training. Some require it at the time of hiring while others require it annually. At a bare minimum, every business should do security awareness training at least once per year. I personally believe you should do training on a monthly basis. Rather than doing a big training that attempts to cover every possible topic in one sitting, host monthly micro trainings that focus on one specific topic.

Studies show that, on average, people remember ten percent of what they read and twenty percent of what they hear. This is why annual large trainings are ineffective. Your people are only retaining a small percentage of the material they're studying in this model, and that assumes the material was easy to understand and follow. Attending monthly micro trainings covering a single topic, employees have a much better chance of retaining the information and keeping cybersecurity top of mind.

At the end of the day, you need to remember one very important thing. The ultimate goal for a criminal is to get to you. "You" being your digital assets like your financials, your intellectual property, your identity, and even your employees' identities. They just use computers to get to you.

Just because you may know what to look out for doesn't mean that everyone within your organization does. Don't become a victim by assuming everyone else is making cybersecurity a priority. Train your people on how to protect themselves so they can do a better job of protecting your business.

PERSONAL LIFE APPLICATION

It's hard for leaders to admit they don't know something. At least that's true for me. Listen, you don't know what you don't know, and that's okay. This was a lesson I learned recently. I used to think that as a leader, husband, and father, I had to have all the answers. Any time I didn't have the answer, I felt the need to fake my way through it. During those moments, I stumbled and rambled through a response rather than admitting that I didn't have the answer. I thought that by admitting I didn't know something, I was diminishing the overall value I brought to the situation.

What I have since learned is that people respect me more when I admit that I don't have the answer but show that I'm committed to finding it. It shows that I have the confidence to be transparent with them. This has taken the pressure off me to feel like I need to be perfect and have the answer. By removing the façade of perfection, I have given myself permission to freely operate in my area of expertise without the pressure of always needing to be right.

Operating at this level of honesty, freedom, and transparency is helping me to be a better leader, communicator, husband, and father. I'm still learning, but I've begun to actively listen during conversations rather than obsessing about how to respond. I can now see the change in the people I communicate with. When I'm actively listening, the other person can see that I'm focusing on them and hearing their point of view. In turn, they're more receptive to what I'm trying to convey. This shift allows for more meaningful and impactful conversations.

I encourage you to be free and live your truth. Your value is not found in your answers but rather in the unique experiences you've had that offer you a different perspective.

CONCLUSION

THE AAR

Perform an annual review using your business optimization blueprint

I was fortunate to have good leadership during my time in the Army. My superiors saw potential in me and they constantly pushed me to become a better thinker, a better soldier, and a better person. The funny thing is, I didn't realize it at the time. As I was going through it, all I could see were the challenges. We were constantly in training to repeat maneuvers until we could complete our tasks at a high level of speed and excellence; or as we commonly referred to it, high speed, low drag.

What really annoyed me were the After Action Reports (AAR). No matter how well things went, we always had to do an AAR to evaluate what went right, what went wrong, why things went wrong, and what we could do in the future to prevent them from going wrong again. Even if nothing went wrong, we performed an AAR. But looking back on it, I understand the purpose of the AAR: it isn't designed to evaluate your results, it's designed to evaluate your process. You look at your processes to see if there is any way to improve. This commitment to constant evaluation allows our military to continue to improve and rise to meet every new challenge presented in the world.

Being a business leader is similar to being a commander in war. I'm obviously not talking about violence. But the overall goals are very similar. In war, you typically have a geographical position you're trying to defend against your adversary. You're also attempting to advance your position without overextending your resources or leaving other areas of your operation vulnerable. And you need to ensure your soldiers have the tools and supplies they need to accomplish their mission.

Does this sound familiar? In business, your adversary is the competition in the market. You're working to either establish a market base or defend your existing market share while expanding and taking over market share from someone else. As you work to expand your business, you have to make sure you don't grow too fast. You can't leave yourself vulnerable, allowing your competition to take your clients due to a decline in the quality of your product/service. You also need to make sure that your employees are properly trained and have the tools they need to complete their job.

This book is not only a tactical guide for optimizing your business; it's also meant to provide you a framework for performing an Annual Action Review (AAR) on your business. You must commit to constantly evaluating and improving your processes so that your business can rise to meet the new challenges in the market. Use the tools you've learned in each chapter to evaluate the three main pillars that will lead to the success of your business: income, efficiency, and defenses. In this concluding chapter, I want to help you take a step back and see how the pieces of this book work together.

Remember that there are three ways to increase your revenue:

1. Sell more of your product/service
2. Increase your rates
3. Do more with less

The tools and processes I've offered you in this book will allow you to accomplish all three. If you maximize the use of technology within your business, you will reduce unnecessary expenses and increase your efficiency, which will allow you to provide your product/service to more clients. If you're able to deliver your product/service to more clients while maintaining a high level of quality, you will increase your rates as the demand for your product/service increases.

Your Business Optimization Blueprint

I. Establish an accurate technology budget so you can see where every dollar spent on technology is going (Tracer Rounds). Maximize the use of every resource available to you by ensuring that you're getting some sort of return or value out of every dollar you spend.

II. Evaluate the operational workflow of your business to discover any bottlenecks, points of friction, or inefficiencies (Triple S method). Prioritize them based on the needs of your business.

III. Create a Technology Roadmap based on the list you made using the Triple S method to govern how and when you invest your technology budget (Sat Nav).

IV. Evaluate the technology solutions that will address the inefficiencies outlined in your Technology Roadmap (STEP into Dominance). This is how you prevent wasteful spending on technology solutions that may sound good but don't support your strategic goals for the business.

V. Seek the guidance of an expert whenever possible (Sat Nav). The knowledge an expert brings to a situation allows you to execute and achieve your desired results in a shorter amount of time; they are well worth the investment. You are paying for speed and accuracy.

VI. Properly train your people (BASIC Training). Your employees need to be well trained in your methods to utilize the tools you have given them. Each member of your staff should embody *your* company culture. Help them continue to grow through ongoing training and education, and you'll build a more balanced team and reduce employee turnover. Getting employees to stick with you for life starts and ends with your company culture.

VII. Set and enforce good policies and procedures (STOP). Your operating procedures establish clear expectations, eliminate confusion, and ensure that employees deliver consistent results.

VIII. Consistently evaluate risk (ERA). As you experience success and your business continues to grow, remember that your risks also grow. Each new person you add to your team and every new technology solution you incorporate into your workflow changes the risk profile of your business. There will also be changes in the market as time goes by that you must be aware of as well.

IX. Minimize risk by educating and training your team to protect themselves so they in turn can protect your business (Fire Guard). Remember that your team members will always be the weakest link in your defense system.

X. Ensure that you can detect the moment something has gone wrong so that you can immediately address it and minimize the damage of cybercrime (Overwatch). As the digital threats we face continue to grow in frequency and complexity, businesses will need to have a dedicated security team to defend against these attacks.

Future-proof Your Business

You can see how all of this works together to create a never-ending process. As your business begins to grow, you will need to evaluate your operational workflow. When you evaluate your operational workflow, you will find opportunities for your team to function more efficiently. As you incorporate new technology solutions and add people to your team to increase your efficiency, you increase your risk profile. Evaluate your risks to make sure you aren't crippled by a threat that you didn't see coming. Implementing these procedures, you'll become more efficient at delivering your product/service, which will allow you to expand your market share and increase revenue, which ultimately starts the cycle all over again. This process of constant execution and evaluation is your AAR. Implementing and sticking to this process is how you develop sustained growth for your business.

This is also how you future-proof your business. Like our ever-evolving military, you can adapt to meet every new challenge and threat in the world. This framework will allow you to objectively evaluate the key pillars of your business that will lead to your success. It doesn't matter if things change within your market or if your company grows so big that it becomes a target, you now have a process that will help you to identify and plan a response to these threats before they become an issue.

If you implement the principles I've laid out for you,

I have no doubt you'll achieve the level of success you desire. Now is the time for you to review your notes and develop a plan. It doesn't matter where you need to start in the process, the most important thing is to get started now while the information is fresh in your mind and you feel motivated to take action.

Finally, remember that success starts in the mind. You have to believe that you are capable of achieving the level of success you desire. That is something I cannot give you. But I will tell you this: many people, including myself, stood in the shadow of big dreams and aspirations and felt the doubt that you may be feeling. Remember that you were created to overcome every challenge you will face along the way and now you have the tools to help you do it.

So start tackling your objectives one by one. I can't wait to see what you're able to accomplish!

AFTERWORD

YANKEE ... GOLF ... DELTA!

Unlock the best version of yourself

When I joined the military, I knew there was a chance I could get deployed. I knew if I was to ever be deployed, there was a chance I could die. This was the hard truth of the matter that I had to face; to be honest, that scared me. The biggest mistake I made back then was thinking that I was the only person who had this fear.

I saw my brothers and sisters in service walking around as if they had no fear or didn't have a care in the world. This was especially true of one of my fellow service members who later became a very dear friend. For the sake of this story, I'll call him James. I was a young guy coming into a world that was new to me. James was already in his eighth year of service. When we met, we instantly hit it off. At least that's how I remember it. If you ask him, he'll tell you he thought I was an arrogant kid in need of guidance. He felt sorry for me, sure that I would be doomed if he didn't take me under his wing.

I've been out of service for almost twenty years now, and James and I still disagree on that. But one thing is for sure—I looked up to James. Nothing fazed him. The guy had ice in his veins. He never backed down and would be the first to tear you a new one if you crossed the line. And yet, he was the kind of guy everyone wanted to be around.

One Saturday afternoon, a group of us were having a cookout. While James and I were trash-talking each other over who could cook better on the grill, I asked him how he did it. How did he stay so cool regardless of the situation? Was it because he was older? Was it because he had more time in service? I wanted to know his secret. I wanted to be like him.

I watched him as the breeze blew the smoke from the grill in my face. I was about to learn the secret to becoming the man I wanted to be.

James took a drink of his beer and said, "Yankee … Golf … Delta."

Completely bewildered, disappointed, and feeling like a child who'd just found out that Santa Claus wasn't real, I asked him what in the hell that meant.

And this is what he said: "Yankee … Golf … Delta. You're Gonna Die. Nobody lives forever. Eventually everyone dies. It is a part of life. The quicker you come to grips with that and stop fearing death, the quicker you can start living life. Who cares what happened yesterday? If you wake up the next morning, then you have a chance to do something new."

At the time, I didn't get it. I thought it was the biggest crock of shit I had ever heard, and we just went on to enjoy the rest of the weekend. As I got older, though, I realized James had actually given me the keys to unlocking my life. I learned two very valuable lessons from him that I want to share with you.

First, let go of fear. The worst thing that can happen to you in life is death. Sure, bad things can happen. You could get hurt. You could lose an arm. You could lose a leg. You could be talked about. You could lose your job or all of your possessions. Imagine a worst-case scenario: whatever it is, it's not worse than death. Death is the absolute end. Game over. There is no redo, no second chance, no reset button. The casket is closed and your ride in this life is over. But death is the fate that we all share. Some of us get to that point fast and some of us get to that point slow. But make no mistake, we will all get to that point at some time. No one knows how much time they have. History has shown us that people die at all ages and in some of the most bizarre ways. Knowing that's the case, why should we be scared of it? It makes no sense to live your life trying to avoid *something that is going to happen regardless of what you do.* When it is your time to go, then it is your time to go. Once I accepted this truth, it was easier for me to start letting go of the other fears that were unknowingly controlling my life.

Second, I gained the freedom to fail. This one was huge for me. I am a very competitive person; I thrive on being right and doing things better than the next person. And while this inner drive and determination helped me to achieve great things, it was also holding me back. I had this image of myself as someone who always had his stuff together, and I was scared to do anything to ruin that image. I hesitated to put myself out there and try new things. I thought if I didn't live up to the ideal, what I now know to be unrealistic expectations, people would think less of me.

"Yankee … Golf … Delta" taught me that as long as I wake up the next day, I have a chance to do something different or better. As long as I am still breathing, nothing is set in stone because I can do something to change it. I can learn something new. I can fix something that I broke. As long as I'm still breathing, I can transform myself into anything I want to be. The only thing that ever has, and ever will, stop me is me. Failing at something does not mean the end. It just means that you tried something and it didn't work out that time. You can assess the situation, learn from your mistakes, and try again, or you can decide that you didn't like it and move on to something else.

The choice is yours. You are the only one who can determine how great your life story turns out to be. I am taking the limits off of your life and giving you permission to fail. Get out in this world and fail. Failing means you're trying new things, and trying new things is part of growing. As we have all seen with COVID-19, nothing is promised. We all came into 2020 with high expectations and hopes for what we might accomplish. No one expected us to lose so many lives to this virus and to see our country and the world thrust into such chaos and turmoil. But I will tell you the same thing I told people at the very beginning of it all: in hard and troubling times, millionaires are made and leaders are born.

As I bring this book to a close, I want to share one more thing with you. I know that you desire to have more in life and on some level you are committed to becoming the best version of yourself. I can say this because you made it to the end of this book. Remember, statistically speaking, ninety-seven percent of people who get a book never read past the seventeenth page. The fact that you have made it this far places you in the top three percent of people and demonstrates the level of dedication you have toward making yourself better.

Where you go from here is completely up to you. You get to choose whether you want to be fully committed and burn the boats, or if you're comfortable with where you are in life. The good news is, there is no wrong choice. No one can tell you what your best life should look like. I have a dear friend who works just so they can live their life out hiking during spring, summer, and fall and skiing all winter. They don't have millions of dollars, but I tell them all the time that they're an inspiration to me. I am inspired because they're living life on their terms, doing the things they love to do. In my eyes, they are living the ultimate dream. Follow your passion, chase your dreams, and live the life you desire. Refuse to settle! You only get one shot at this thing called life. You might as well get out there and experience all it has to offer and create your legacy.

My desire to create my legacy led me to write this book. By sharing my experience and expertise with you, I hope to support you in achieving your goals and encourage you to start living the life you desire. If only one part of this book helps you grow your business or encourages you to start taking action toward living your best life right now, then it was all worth it to me. All I ask is that you pay it forward. Share the message with others. Inspire others through your story and encourage others to do the same. One by one, we can impact lives and create positive change that will impact generations to come. And that, my friend, is a legacy we can all be proud of.

Thanks again for the support. I wish you nothing but continued success, and I look forward to the day that our paths may cross again. May the best of your todays be the worst of your tomorrows. Take care!

To God be the Glory,

A-Jay Orr

ACKNOWLEDGMENTS

Writing this book was harder than I thought it would be. But it was also more rewarding than I could have ever imagined. First, I want to praise God for giving me the strength and courage to put myself out there. Without Him, none of this would have been possible.

I could not have done this without the unconditional love and support of my best friend and wife, Amy. No matter what is going on, you are always my biggest cheerleader, my biggest fan. You are the support that holds me up. I would not be the man I am if it wasn't for your constant prayers and your refusal to give up on me. Thank you for sticking by me and giving me the freedom to do what I do.

To my kids, Xander, Vivian, Penelope, and Cooper—you are my why. You are the reason I push myself and work so hard. My desire has always been to blaze a trail for you. I want to be the type of father you can look up to and be proud of. I hope I can continue to inspire you and show you that anything is possible as long as you're willing to take the limits off of your dreams and put in the work to achieve them. Above all else, know that I will always be here to catch you when you stumble. Chase your dreams with all you have because only you have the power to make them come true.

To my loving parents, Augustus and Sharon Orr, thank you for your support and guidance through the years. I know I was a challenging child who pushed you to your limits and worked every nerve you had, but you always made me feel loved.

I would be remiss if I didn't show my appreciation to the United States Army. You took a young, undisciplined, unfocused, cocky, arrogant, and hotheaded kid and turned him into a focused and disciplined man capable of doing anything he sets his mind to. Not only did you save me from myself, but you also gave me countless brothers and sisters who will be there to pick me up if ever I should fall. Thank you.

Finally, to every person I have had the pleasure of interacting with throughout my life, I want to say thank you. Whether the experience was good or bad, that interaction played a part in shaping me into the person I am.

THANK YOU

Thank you so much for reading! I truly hope you enjoyed it. For more content like this, visit www.ajayorr.com to learn how to work with me or schedule me for a speaking engagement. You can also connect with me on social media:

LinkedIn: https://www.linkedin.com/in/ajayorr

Twitter: https://twitter.com/ajayorr

Facebook: https://www.facebook.com/iamajayorr

Instagram: https://www.instagram.com/iamajayorr

Thanks so much for choosing to go on this journey with me. Authors like me really need and appreciate support from folks like you, and I'm glad you stopped by. Please don't hesitate to connect with me if you have any questions about the book. I enjoy connecting with readers and would love to hear from you!

A-Jay

A QUICK FAVOR PLEASE?

Before you go, I have a quick favor… Would you please leave this book a review? Thoughtful reviews are especially important for authors like me, as they help me reach more readers like you and, in turn, enable me to continue publishing great content. It doesn't take long, and I genuinely appreciate your time and support.

Thanks again,

–A-Jay

ABOUT THE AUTHOR

A-Jay Orr is the founder and CEO of Simple Plan IT, an internationally certified project manager and global digital risk and cybersecurity expert with more than fifteen years of experience in this ever-evolving field. He is the expert that companies rely on to provide sophisticated solutions to achieve business growth and maintain compliance. He implements behavior-based security solutions to protect manufacturers and small businesses from cyberattacks that target control systems, robotics, IoT, and smart devices that often go undetected. A-Jay proudly served his country and is a veteran of the United States Army. He also gives back to the community as a Bunker Labs Ohio Ambassador, where he helps veteran and veteran spouses to become successful entrepreneurs and business owners.

A-Jay is a husband to his beautiful wife, Amy, and is the father of four amazing kids, Xander, Vivian, Penelope, and Cooper. He also enjoys traveling and is a diehard fan of Ohio State Buckeyes football, Kentucky Wildcats basketball, the Los Angeles Lakers, and the Dallas Cowboys.

Made in United States
Orlando, FL
10 November 2026

90487622R10085